Autonomous Mobile Robots

Autonomous Mobile Robots

Yogesh Kumar

RANDOM
RANDOM PUBLICATIONS
NEW DELHI (INDIA)

Autonomous Mobile Robots

ISBN 978-93-5111-236-5

Published in 2014 in India by

RANDOM PUBLICATIONS

4376-A/4B, Gali Murari Lal, Ansari Road
New Delhi-110 002
Phone : +91-11-43580356, +91-11-23289044
e-mail: randomexports@gmail.com, sales@randompublications.com, info@randompublications.com

Reprinted 2026

Type Setting by : Keystoneprintads, Delhi-110051

Preface

Autonomous mobile robots (or agents) require a quite different way of thinking. The autonomous-agents perspective is not so much on isolated computational issues, such as choosing a specific neural network or fuzzy rule system, but rather on the agent-environment interaction.

Mobile robots can support humans in various ways. They operate in hazardous and inaccessible areas and make work easier for people by relieving them of monotonous tasks. Current applications range from assistance robots within the industrial environment to reconnaissance robots in disaster areas and to robotic systems for deep-sea inspection and exploration tasks. The demands placed on the functionality of mobile robotic systems are significantly higher compared to conventional industrial robots. The aim is that mobile systems operate autonomously in unknown and dynamic environments to fulfill their assigned tasks. For this purpose, it is essential to explore and model the environment in a suitable way. The information gathered by sensors must be combined to allow for an accurate positioning. In addition, the perceived surroundings must be consolidated in an exact map representation. Having acquired this knowledge, the robot must be able to plan an optimal collision-free path to a given goal.

Mobile robots are deployed in dangerous or inaccessible environments and can support human beings in numerous ways. The algorithm toolbox allows a wide variety of mobile platforms to be equipped with the autonomous skills relevant for a specific purpose without any need for major customization. The functional requirements for mobile robot systems are considerably higher than those for conventional industrial robots.

Mobile systems are supposed to act autonomously in an unknown, dynamic environment and fulfill a technical function. To this end, it is indispensable to explore the environment and to model it in an appropriate way. The information collected by sensors has to be coordinated, allowing the robot to determine its exact position.

The book presents the techniques and technology that enable mobility in a series of interacting modules. Each chapter treats a different aspect of mobility, as the book moves from low-level to high-level details. It covers all aspects of mobile robotics, including software and hardware design considerations, related technologies, and algorithmic techniques.

I thank all members of my team who have helped in the preparation of the book. My special thanks go to "Random Publications" who have published the book.

– ***Yogesh Kumar***

Contents

1

Introduction to Mobile Robot

MOBILE ROBOT

A mobile robot is an automatic machine that is capable of movement in any given environment.

Fig. A spying robot is an example of a mobile robot capable of movement in a given environment.

Mobile robots have the capability to move around in their environment and are not fixed to one physical location. In contrast,industrial robots usually consist of a jointed arm (multi-linked manipulator) and gripper assembly (or end effector) that is attached to a fixed surface.

Mobile robots are a major focus of current research and almost every major university has one or more labs that focus on mobile robot research. Mobile robots are also found in industry, military and security environments. Domestic robots are consumer products, including entertainment robots and those that perform certain household tasks such as vacuuming or gardening.

CLASSIFICATION

Mobile robots may be classified by:

The environment in which they travel:

- Land or home robots are usually referred to as Unmanned Ground

Vehicles (UGVs). They are most commonly wheeled or tracked, but also include legged robots with two or more legs (humanoid, or resembling animals or insects).

- Aerial robots are usually referred to as Unmanned Aerial Vehicles (UAVs)
- Underwater robots are usually called autonomous underwater vehicles (AUVs)
- Polar robots, designed to navigate icy, crevasse filled environments

The device they use to move, mainly:

- Legged robot: human-like legs (i.e. an android) or animal-like legs.
- Wheeled robot.
- Tracks.

MOBILE ROBOT NAVIGATION

There are many types of mobile robot navigation:

Manual remote or tele-op

A manually teleoperated robot is totally under control of a driver with a joystick or other control device. The device may be plugged directly into the robot, may be a wireless joystick, or may be an accessory to a wireless computer or other controller. A tele-op'd robot is typically used to keep the operator out of harm's way. Examples of manual remote robots include Robotics Design's ANATROLLER ARI-100 and ARI-50, Foster-Miller's Talon, iRobot's PackBot, and KumoTek's MK-705 Roosterbot.

Guarded tele-op

A guarded tele-op robot has the ability to sense and avoid obstacles but will otherwise navigate as driven, like a robot under manual tele-op. Few if any mobile robots offer only guarded tele-op.

Line-following robot

Some of the earliest Automated Guided Vehicles (AGVs) were line following mobile robots. They might follow a visual line painted or embedded in the floor or ceiling or an electrical wire in the floor.

Most of these robots operated a simple "keep the line in the center sensor" algorithm. They could not circumnavigate obstacles; they just stopped and waited when something blocked their path. Many examples of such vehicles are still sold, by Transbotics, FMC, Egemin, HK Systems and many other companies.

Autonomously randomized robot

Autonomous robots with random motion basically bounce off walls, whether those walls are sensed

Autonomously guided robot

Fig. Robot developers use ready-made autonomous bases and software to design robot applications quickly. Shells shaped like people or cartoon characters may cover the base to disguise it. Courtesy of MobileRobots Inc

An autonomously guided robot knows at least some information about where it is and how to reach various goals and or waypoints along the way. "Localization" or knowledge of its current location, is calculated by one or more means, using sensors such motor encoders, vision, Stereopsis, lasers and global positioning systems.

Positioning systems often use triangulation, relative position and/or Monte-Carlo/Markov localization to determine the location and orientation of the platform, from which it can plan a path to its next waypoint or goal. It can gather sensor readings that are time- and location-stamped, so that a hospital, for instance, can know exactly when and where radiation levels exceeded permissible levels. Such robots are often part of the wireless enterprise network, interfaced with other sensing and control systems in the building.

For instance, the PatrolBot security robot responds to alarms, operates elevators and notifies the command center when an incident arises. Other autonomously guided robots include the SpeciMinder and the Tug delivery robots for hospital labs, though the latter actually has people at the ready to drive the robot remotely when its autonomy fails. The Tug sends a letter to its tech support person, who then takes the helm and steers it over the Internet by looking through a camera low in the base of the robot.

Sliding autonomy

More capable robots combine multiple levels of navigation under a system called sliding autonomy. Most autonomously guided robots, such as the HelpMate hospital robot, also offer a manual mode. The Motivity autonomous robot operating system, which is used in the ADAM, PatrolBot, SpeciMinder, MapperBot and a number of other robots, offers full sliding autonomy, from manual to guarded to autonomous modes.

HISTORY

Date	Developments
1939–1945	During World War II the first mobile robots emerged as a result of technical advances on a number of relatively new research fields like computer science andcybernetics. They were mostly flying bombs. Examples are smart bombs that only detonate within a certain range of the target, the use of guiding systems and radar control. The V1 and V2 rockets had a crude 'autopilot' and automatic detonation systems. They were the predecessors of modern cruise missiles.
1948–1949	W. Grey Walter builds Elmer and Elsie, two autonomous robots called Machina Speculatrix because these robots liked to explore their environment. Elmer and Elsie were each equipped with a light sensor. If they found a light source they would move towards it, avoiding or moving obstacles on their way. These robotsdemonstrated that complex behaviour could arise from a simple design. Elmer and Elsie only had the equivalent of two nerve cells.
1961–1963	The Johns Hopkins University develops 'Beast'. Beast used a sonar to move around. When its batteries ran low it would find a power socket and plug itself in.
1969	Mowbot was the very first robot that would automatically mow the lawn.
1970	The Stanford Cart line follower was a mobile robot that was able to follow a white line, using a camera to see. It was radio linked to a large mainframe that made the calculations. At about the same time (1966–1972) the Stanford Research Institute is building and doing research on Shakey the Robot, a robot named after its jerky motion. Shakey had a camera, a rangefinder, bump sensors and a radio link. Shakey was the first robot that could reason about its actions. This means that Shakeycould be given very general commands, and that the robot would figure out the necessary steps to accomplish the given task. The Soviet Union explores the surface of the Moon with Lunokhod 1, a lunar rover.
1976	In its Viking program the NASA sends two unmanned spacecraft to Mars.

1980	The interest of the public in robots rises, resulting in robots that could be purchased for home use. These robots served entertainment or educational purposes. Examples include the RB5X, which still exists today and the HERO series. The Stanford Cart is now able to navigate its way through obstacle courses and make maps of its environment.
Early 1980s	The team of Ernst Dickmanns at Bundeswehr University Munich builds the first robot cars, driving up to 55 mph on empty streets.
1987	Hughes Research Laboratories demonstrates the first cross-country map and sensor-based autonomous operation of a robotic vehicle.
1989	Mark Tilden invents BEAM robotics.
1990s	Joseph Engelberger, father of the industrial robotic arm, works with colleagues to design the first commercially available autonomous mobile hospital robots, sold by Helpmate. The US Department of Defense funds the MDARS-I project, based on the Cybermotion indoor security robot.
1991	Edo. Franzi, André Guignard and Francesco Mondada developed Khepera, an autonomous small mobile robot intended for research activities. The project was supported by the LAMI-EPFL lab.
1993–1994	Dante I and Dante II were developed by Carnegie Mellon University. Both were walking robots used to explore live volcanoes.
1994	With guests on board, the twin robot vehicles VaMP and VITA-2 of Daimler-Benz and Ernst Dickmanns of UniBwM drive more than one thousand kilometers on a Paris three-lane highway in standard heavy traffic at speeds up to 130 km/h. They demonstrate autonomous driving in free lanes, convoy driving, and lane changes left and right with autonomous passing of other cars.
1995	Semi-autonomous ALVINN steered a car coast-to-coast under computer control for all but about 50 of the 2850 miles. Throttle and brakes, however, were controlled by a human driver.
1995	In the same year, one of Ernst Dickmanns' robot cars (with robot- controlled throttle and brakes) drove more than 1000 miles from Munich to Copenhagen and back, in traffic, at up to 120 mph, occasionally executing maneuvers to pass other cars (only in a few critical situations a safety driver took over). Active vision was used to deal with rapidly changing street scenes.
1995	The Pioneer programmable mobile robot becomes commercially

	available at an affordable price, enabling a widespread increase in robotics research and university study over the next decade as mobile robotics becomes a standard part of the university curriculum.
1996–1997	NASA sends the Mars Pathfinder with its rover Sojourner to Mars. The rover explores the surface, commanded from earth. Sojourner was equipped with a hazard avoidance system. This enabled Sojourner to autonomously find it s way through unknown martian terrain.
1999	Sony introduces Aibo, a robotic dog capable of seeing, walking and interacting with its environment. The PackBot remote-controlled military mobile robot is introduced.
2001	Start of the Swarm-bots project. Swarm bots resemble insect colonies. Typically they consist of a large number of individual simple robots, that can interact with each other and together perform complex tasks. [4]
2002	Appears Roomba, a domestic autonomous mobile robot that cleans the floor.
2003	Axxon Robotics purchases Intellibot, manufacturer of a line of commercial robots that scrub, vacuum, and sweep floors in hospitals, office buildings and other commercial buildings. Floor care robots from Intellibot Robotics LLC operate completely autonomously, mapping their environment and using an array of sensors for navigation an obstacle avoidance.
2004	Robosapien, a biomorphic toy robot designed by Mark Tilden is commercially available. In 'The Centibots Project' 100 autonomous robots work together to make a map of an unknown environment and search for objects within the environment. In the first DARPA Grand Challenge competition, fully autonomous vehicles compete against each other on a desert course.
2005	Boston Dynamics creates a quadruped robot intended to carry heavy loads across terrain too rough for vehicles.
2006	Sony stops making Aibo and HelpMate halts production, but a lower-cost PatrolBot customizable autonomous service robot system becomes available as mobile robots continue the struggle to become commercially viable. The US Department of Defense drops the MDARS-I project, but funds MDARS-E, an autonomous field robot. TALON-Sword, the first commercially available robot with grenade launcher and other integrated weapons options, is released. Honda's Asimo learns to run and climb stairs.

2007	In the DARPA Urban Grand Challenge, six vehicles autonomously comple a complex course involving manned vehicles and obstacles. Kiva Systems clever robots proliferate in distribution operations; these smart shelving units sort themselves according to the popularity of their contents. The Tug becomes a popular means for hospitals to move large cabinets of stock from place to place, while the Speci-Minder [5] with Motivity begins carrying blood and other patient samples from nurses' stations to various labs. Seekur, the first widely available, non-military outdoor service robot, pulls a 3-ton vehicle across a parking lot,drives autonomously indoors and begins learning how to navigate itself outside. Meanwhile, PatrolBot learns to follow people and detect doors that are ajar.
2008	Boston Dynamics released video footage of a new generation BigDog able to walk on icy terrain and recover its balance when kicked from the side.
2010	The Multi Autonomous Ground-robotic International Challenge has teams of autonomous vehicles map a large dynamic urban environment, identify and track humans and avoid hostile objects.

DOMESTIC ROBOT

Fig. First generation Roomba vacuums the carpets in a domestic environment

A domestic robot, or service robot, is an autonomous robot that is used for household chores. Thus far, there are only a few limited models, though speculators, such as Bill Gates, have suggested that they could become more common in the future.Many domestic robots are used for basic household chores. Others are educational or entertainment robots, such as the HEROline of the 1980s. While most domestic robots are simplistic, some are connected to WiFi home networks or smart environmentsand are autonomous to a high degree. There were an estimated 3,540,000 service robots in use in 2006, compared with an estimated 950,000 industrial robots.

IN-HOME ROBOTS

This type of domestic robot does chores around and inside homes. Different kinds include:

Robotic vacuum cleaners and floor-washing robots that clean floors with sweeping and wet mopping functions. Some use Swiffer or other disposable cleaning cloths to dry-sweep, or reusable microfiber cloths to wet-mop.

Cat litter robots are automatic self-cleaning litter boxes that filter clumps out into a built-in waste receptacle that can be lined with an ordinary plastic bag. Security robots which have a night-vision-capable wide-angle camera that detects movements and intruders. It can patrol places and shoot video of suspicious activities, too, and send alerts via email or text message; the stored history of past alerts and videos are accessible via the Web. The robot can also be configured to go into action at any time of the day.

OUTDOOR ROBOTS

Outdoor robots are domestic robots that perform different chores that exist outside of the house. Robotic lawn mowers are one type of outdoor robot that cut grass on their own without the need for a driver. Some models can mow complicated and uneven lawns that are up to three-quarters of an acre in size. Others can mow a lawn as large as 40,000 square feet, can handle a hill inclined up to 27 degrees.

There are also automated pool cleaners that clean and maintain swimming pools autonomously by scrubbing in-ground pools from the floor to the waterline in 3 hours, cleaning and circulating more than 70 gallons of water per minute, and removing debris as small as 2 microns (2 millionths of a meter) in size. Window-washing robots use two magnetic modules to navigate your windows as it sprays cleaning solution onto microfiber pads to wash them. It covers about 1601 square feet per charge. Gutter-cleaning robots can blast through debris, clogs, and sludge in gutters, and brushes them clean.

TOYS

Robotic toys, such as the well known Furby, have been popular since 1998. There are also small humanoid remote controlled robots. Electronic pets, such as robotic dogs, can be companions for children. They have also have been used by many universities in competitions such as the RoboCup.

There are also phone-powered robots for fun and games, such as Romo which is a small robot that employs smartphones as its brain. By using another mobile device and a cross-platform app, the user can drive it, make it produce animated facial expressions, direct it to dance, or turn it into a spybot.

SOCIAL ROBOTS

A social robot are robots whose main objective is social interaction. A lot of these robots are designed to help the elderly. For example, the Wakamaru is a humanoid robot designed to provide company for elderly and less mobile people, made by Mitsubishi Heavy Industries.

There is also the Paro, a robot baby seal intended to provide comfort to nursing home patients.

Home-telepresence robots can move around in a remote location and let you communicate with people there via its camera, speaker, and microphone. Through other remote-controlled telepresence robots, anyone can visit a distant location and explore it as if they were physically present. It lets health-care workers check on patients easily, for instance, and kids who are homebound because of injuries, illnesses, or physical challenges can go to school via their VGo.

ROBOTS NO LONGER IN PRODUCTION

Early historical attempts to bring robots into the home. Their actual functions is up for debate.

- Hero, 1982–89
- Topo, 1983

IN POPULAR CULTURE

Many cartoons feature robot maids, notably Rosie the Robot from The Jetsons. Maid Robots are especially prominent in anime (in Japanese, they are called Meido Robo or Meido Roboto), and their Artificial Intelligence ranges from rudimentary to fully sentient and emotional, while their appearance ranges from obviously mechanical to human-like.

The 2012 movie Robot & Frank featured a domestic robot, the story of the movie centred around an elderly man and his relationship with a caretaker robot.

The 2005 movie Robots is a children's film, about a young inventor in a robot world traveling the big city to join his inspiration's company.

HUMANOID ROBOT

A humanoid robot is a robot with its body shape built to resemble that of the human body. A humanoid design might be for functional purposes, such as interacting with human tools and environments, for experimental purposes, such as the study of bipedal locomotion, or for other purposes. In general, humanoid robots have a torso, a head, two arms, and two legs, though some forms of humanoid robots may model only part of the body, for example, from the waist up.

Some humanoid robots may also have heads designed to replicate human facial features such as eyes and mouths. Androids are humanoid robots built to aesthetically resemble humans.

PURPOSE

Fig. TOPIO, a humanoid robot, playedping pong at Tokyo International Robot Exhibition (IREX) 2009.

Fig. Nao (robot) is a robot created for companionship. It also competes in the RoboCupsoccer championship.

Humanoid robots are used as a research tool in several scientific area.

Researchers need to understand the human body structure and behavior (biomechanics) to build and study humanoid robots. On the other side, the attempt to the simulation of the human body leads to a better understanding of it.

Human cognition is a field of study which is focused on how humans learn from sensory information in order to acquire perceptual and motor skills. This knowledge is used to develop computational models of human behavior and it has been improving over time.

It has been suggested that very advanced robotics will facilitate the enhancement of ordinary humans. See transhumanism.

Although the initial aim of humanoid research was to build better orthosis and prosthesis for human beings, knowledge has been transferred between both disciplines. A few examples are: powered leg prosthesis for neuromuscularly impaired, ankle-foot orthosis, biological realistic leg prosthesis and forearm prosthesis.

Besides the research, humanoid robots are being developed to perform human tasks like personal assistance, where they should be able to assist the sick and elderly, and dirty or dangerous jobs. Regular jobs like being a receptionist or a worker of an automotive manufacturing line are also suitable for humanoids. In essence, since they can use tools and operate equipment and vehicles designed for the human form, humanoids could theoretically perform any task a human being can, so long as they have the proper software. However, the complexity of doing so is deceptively great.

They are becoming increasingly popular for providing entertainment too. For example, Ursula, a female robot, sings, play music, dances, and speaks to her audiences at Universal Studios. Several Disney attractions employ the use of animatrons, robots that look, move, and speak much like human beings, in some of their theme park shows. These animatrons look so realistic that it can be hard to decipher from a distance whether or not they are actually human. Although they have a realistic look, they have no cognition or physical autonomy. Various humanoid robots and their possible applications in daily life are featured in an independent documentary film called Plug & Pray, which was released in 2010.

Humanoid robots, especially with artificial intelligence algorithms, could be useful for future dangerous and/or distant space explorationmissions, without having the need to turn back around again and return to Earth once the mission is completed.

SENSORS

A sensor is a device that measures some attribute of the world. Being one of the three primitives of robotics (besides planning and control), sensing plays an important role in robotic paradigms. Sensors can be classified according to the physical process with which they work or according to the type of measurement information that they give as output. In this case, the second approach was used.

Proprioceptive Sensors

Proprioceptive sensors sense the position, the orientation and the speed of the humanoid's body and joints. In human beings inner ears are used to maintain balance and orientation. Humanoid robots use accelerometers to measure the acceleration, from which velocity can be calculated by integration; tilt sensors to measure inclination; force sensors placed in robot's hands and feet to measure contact force with environment; position sensors, that indicate the actual position of the robot (from which the velocity can be calculated by derivation) or even speed sensors.

Exteroceptive Sensors

Arrays of tactels can be used to provide data on what has been touched. The Shadow Hand uses an array of 34 tactels arranged beneath its polyurethane skin on each finger tip. Tactile sensors also provide information about forces and torques transferred between the robot and other objects. Vision refers to processing data from any modality which uses the electromagnetic spectrum to produce an image. In humanoid robots it is used to recognize objects and determine their properties. Vision sensors work most similarly to the eyes of human beings. Most humanoid robots use CCD cameras as vision sensors.

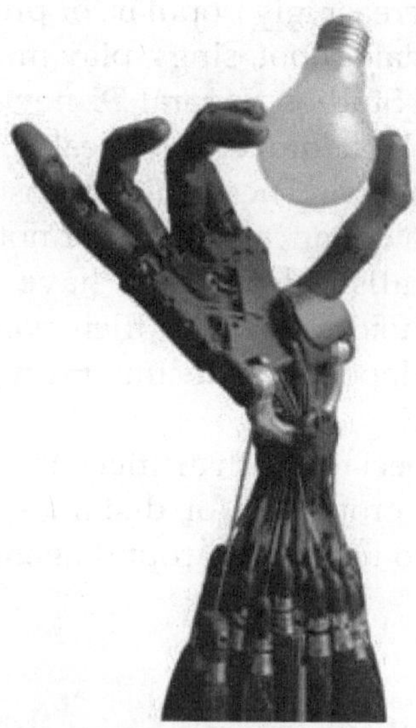

Fig. An artificial hand holding a lightbulb

Sound sensors allow humanoid robots to hear speech and environmental sounds, and perform as the ears of the human being.Microphones are usually used for this task.

ACTUATORS

Actuators are the motors responsible for motion in the robot. Humanoid robots are constructed in such a way that they mimic the human body, so they use actuators that perform like muscles andjoints, though with a different structure. To achieve the same effect as human motion, humanoid robots use mainly rotary actuators. They can be either electric, pneumatic, hydraulic, piezoelectric or ultrasonic.

Hydraulic and electric actuators have a very rigid behavior and can only be made to act in a compliant manner through the use of relatively complex feedback control strategies. While electric coreless motor actuators are better suited for high speed and low load applications, hydraulic ones operate well at low speed and high load applications. Piezoelectric actuators generate a small movement with a high force capability when voltage is applied. They can be used for ultra-precise positioning and for generating and handling high forces or pressures in static or dynamic situations.

Ultrasonic actuators are designed to produce movements in a micrometer order at ultrasonic frequencies (over 20 kHz). They are useful for controlling vibration, positioning applications and quick switching. Pneumatic actuators operate on the basis of gas compressibility. As they are inflated, they expand along the axis, and as they deflate, they contract. If one end is fixed, the other will move in a linear trajectory. These actuators are intended for low speed and low/medium load applications. Between pneumatic actuators there are: cylinders,bellows, pneumatic engines, pneumatic stepper motors and pneumatic artificial muscles.

PLANNING AND CONTROL

In planning and control, the essential difference between humanoids and other kinds of robots (like industrial ones) is that the movement of the robot has to be human-like, using legged locomotion, especially biped gait. The ideal planning for humanoid movements during normal walking should result in minimum energy consumption, like it does in the human body. For this reason, studies on dynamics and control of these kinds of structures become more and more important. To maintain dynamic balance during the walk, a robot needs information about contact force and its current and desired motion. The solution to this problem relies on a major concept, the Zero Moment Point (ZMP).

Another characteristic of humanoid robots is that they move, gather information (using sensors) on the "real world" and interact with it. They don't stay still like factory manipulators and other robots that work in highly structured environments. To allow humanoids to move in complex environments, planning and control must focus on self-collision detection, path planning and obstacle avoidance.

Humanoids don't yet have some features of the human body. They include structures with variable flexibility, which provide safety (to the robot itself and to the people), and redundancy of movements, i.e. more degrees of freedom and therefore wide task availability. Although these characteristics are desirable to humanoid robots, they will bring more complexity and new problems to planning and control.

Timeline of developments

Year	Development
c. 250 BC	The Lie Zi described an automaton.
c. 50 AD	Greek mathematician Hero of Alexandria described a machine to automatically pour wine for party guests.
1206	Al-Jazari described a band made up of humanoid automata which, according to Charles B. Fowler, performed "more than fifty facial and body actions during each musical selection." Al-Jazari also

created hand washing automata with automatic humanoid servants, and an elephant clock incorporating an automatic humanoid mahout striking a cymbal on the half-hour. His programmable "castle clock" also featured five musician automata which automatically played music when moved by levers operated by a hidden camshaft attached to a water wheel.

1495 Leonardo da Vinci designs a humanoid automaton that looks like an armored knight, known as Leonardo's robot.

1738 Jacques de Vaucanson builds The Flute Player, a life-size figure of a shepherd that could play twelve songs on the flute and The Tambourine Player that played a flute and a drum or tambourine.

1774 Pierre Jacquet-Droz and his son Henri-Louis created the Draughtsman, the Musicienne and the Writer, a figure of a boy that could write messages up to 40 characters long.

1898 Nikola Tesla publicly demonstrates his "automaton" technology by wirelessly controlling a model boat at the Electrical Exposition held at Madison Square Garden in New York City during the height of the Spanish-American War.

1921 Czech writer Karel Capek introduced the word "robot" in his play R.U.R. (Rossum's Universal Robots). The word "robot" comes from the word "robota", meaning, in Czech and Polish, "forced labour, drudgery".

1927 The Maschinenmensch ("machine-human"), a gynoid humanoid robot, also called "Parody", "Futura", "Robotrix", or the "Maria impersonator" (played by German actress Brigitte Helm), perhaps the most memorable humanoid robot ever to appear on film, is depicted in Fritz Lang's film Metropolis.

1941-42 Isaac Asimov formulates the Three Laws of Robotics, and in the process of doing so, coins the word "robotics".

1948 Norbert Wiener formulates the principles of cybernetics, the basis of practical robotics.

1961 The first digitally operated and programmable non-humanoid robot, the Unimate, is installed on a General Motors assembly line to lift hot pieces of metal from a die casting machine and stack them. It was created by George Devol and constructed by Unimation, the first robot manufacturing company.

1969 D.E. Whitney publishes his article "Resolved motion rate control of manipulators and human prosthesis".

1970 Miomir Vukobratovic has proposed Zero Moment Point, a theoretical model to explain biped locomotion.

1972 Miomir Vukobratovic and his associates at Mihajlo Pupin Institute build the first active anthropomorphic exoskeleton.

1973 In Waseda University, in Tokyo, Wabot-1 is built. It was able to walk, to communicate with a person in Japanese and to measure distances and directions to the objects using external receptors, artificial ears and eyes, and an artificial mouth.

1980 Marc Raibert established the MIT Leg Lab, which is dedicated to studying legged locomotion and building dynamic legged robots.

1983 Using MB Associates arms, "Greenman" was developed by Space and Naval Warfare Systems Center, San Diego. It had an exoskeletal master controller with kinematic equivalency and spatial correspondence of the torso, arms, and head. Its vision system consisted of two 525-line video cameras each having a 35-degree field of view and video camera eyepiece monitors mounted in an aviator's helmet.

1984 At Waseda University, the Wabot-2 is created, a musician humanoid robot able to communicate with a person, read a normal musical score with his eyes and play tunes of average difficulty on an electronic organ.

1985 Developed by Hitachi Ltd, WHL-11 is a biped robot capable of static walking on a flat surface at 13 seconds per step and it can also turn.

1985 WASUBOT is another musician robot from Waseda University. It performed a concerto with the NHK Symphony Orchestra at the opening ceremony of the International Science and Technology Exposition.

1986 Honda developed seven biped robots which were designated E0 (Experimental Model 0) through E6. E0 was in 1986, E1 – E3 were done between 1987 and 1991, and E4 - E6 were done between 1991 and 1993.

1989 Manny was a full-scale anthropomorphic robot with 42 degrees of freedom developed at Battelle's Pacific Northwest Laboratories in Richland, Washington, for the US Army's Dugway Proving Ground in Utah. It could not walk on its own but it could crawl, and had an artificial respiratory system to simulate breathing and sweating.

1990 Tad McGeer showed that a biped mechanical structure with knees could walk passively down a sloping surface.

1993 Honda developed P1 (Prototype Model 1) through P3, an evolution from E series, with upper limbs. Developed until 1997.

1995 Hadaly was developed in Waseda University to study human-robot communication and has three subsystems: a head-eye subsystem, a voice control system for listening and speaking in Japanese, and a motion-control subsystem to use the arms to point toward campus destinations.

1995 Wabian is a human-size biped walking robot from Waseda University.

1996 Saika, a light-weight, human-size and low-cost humanoid robot, was developed at Tokyo University. Saika has a two-DOF neck, dual five-DOF upper arms, a torso and a head. Several types of hands and forearms are under development also. Developed until 1998.

1997 Hadaly-2, developed at Waseda University, is a humanoid robot which realizes interactive communication with humans. It communicates not only informationally, but also physically.

2000 Honda creates its 11th bipedal humanoid robot, able to run, ASIMO.

2001 Sony unveils small humanoid entertainment robots, dubbed Sony Dream Robot (SDR). Renamed Qrio in 2003.

2001 Fujitsu realized its first commercial humanoid robot named HOAP-1. Its successors HOAP-2 and HOAP-3 were announced in 2003 and 2005, respectively. HOAP is designed for a broad range of applications for R&D of robot technologies.

2002 HRP-2, biped walking robot built by the Manufacturing Science and Technology Center (MSTC) in Tokyo.

2003 JOHNNIE, an autonomous biped walking robot built at the Technical University of Munich. The main objective was to realize an anthropomorphic walking machine with a human-like, dynamically stable gait

2003 Actroid, a robot with realistic silicone "skin" developed by Osaka University in conjunction with Kokoro Company Ltd.

2004 Persia, Iran's first humanoid robot, was developed using realistic simulation by researchers of Isfahan University of Technology in conjunction with ISTT.

2004 KHR-1, a programmable bipedal humanoid robot introduced in June 2004 by a Japanese company Kondo Kagaku.

2005 The PKD Android, a conversational humanoid robot made in the likeness of science fiction novelist Philip K Dick, was developed as a collaboration between Hanson Robotics, the FedEx Institute of Technology, and the University of Memphis.

2005 Wakamaru, a Japanese domestic robot made by Mitsubishi Heavy Industries, primarily intended to provide companionship to elderly and disabled people.

2005 Nao is a small open source programmable humanoid robot developed by Aldebaran Robotics, in France. Widely used by world wide universities as a research platform and educational tool.

2005 The Geminoid series is a series of ultra-realistic humanoid robots or Actroid developed by Hiroshi Ishiguro of ATR and Kokoro in Tokyo. The original one, Geminoid HI-1 was made at its image.

Followed Geminoid-F in 2010 and Geminoid-DK in 2011.

2006 RoboTurk is designed and realized by Dr Davut Akdas and Dr Sabri Bicakci at Balikesir University. This Research Project Sponsored By The Scientific And Technological Research Council Of Turkey (TUBITAK) in 2006. RoboTurk is successor of biped robots named "Salford Lady" and "Gonzalez" at university of Salford in the UK. It is the first humanoid robot supported by Turkish Government.

2006 REEM-A, a biped humanoid robot designed to play chess with the Hydra Chess engine. The first robot developed by PAL Robotics, it was also used as a walking, manipulation speech and vision development platform.

2006 iCub, a biped humanoid open source robot for cognition research.

2006 Mahru, a network-based biped humanoid robot developed in South Korea.

2007 TOPIO, a ping pong playing robot developed by TOSY Robotics JSC.

2007 Twendy-One, a robot developed by the WASEDA University Sugano Laboratory for home assistance services. It is not biped, as it uses an omni-directional mobile mechanism.

2008 Justin, a humanoid robot developed by the German Aerospace Center (DLR).

2008 KT-X, the first international humanoid robot developed as a collaboration between the five-time consecutive RoboCup champions, Team Osaka, and KumoTek Robotics.

2008 Nexi, the first mobile, dexterous and social robot, makes its public debut as one of TIME magazine's top inventions of the year. The robot was built through a collaboration between the MIT Media Lab Personal Robots Group, UMass Amherst and Meka robotics.

2008 Salvius (robot), The first open source humanoid robot built in the United States is created.

2008 REEM-B, the second biped humanoid robot developed by PAL Robotics. It has the ability to autonomously learn its environment using various sensors and carry 20% of its own weight.

2009 HRP-4C, a Japanese domestic robot made by National Institute of Advanced Industrial Science and Technology, shows human characteristics in addition to bipedal walking.

2009 Turkey's first dynamically walking humanoid robot, SURALP, is developed by Sabanci University in conjunction with Tubitak.

2009 Kobian, an robot developed by WASEDA University can walk, talk and mimic emotions.

2010 NASA and General Motors revealed Robonaut 2, a very advanced humanoid robot. It was part of the payload of Shuttle Discovery on

	the successful launch February 24, 2011. It is intended to do spacewalks for NASA.
2010	Students at the University of Tehran, Iran unveil the Surena II. It was unveiled by President Mahmoud Ahmadinejad.
2010	Researchers at Japan's National Institute of Advanced Industrial Science and Technology demonstrate their humanoid robot HRP-4C singing and dancing along with human dancers.
2010	In September the National Institute of Advanced Industrial Science and Technology also demonstrates the humanoid robot HRP-4. The HRP-4 resembles the HRP-4C in some regards but is called "athletic" and is not a gynoid.
2010	REEM, a humanoid service robot with a wheeled mobile base. Developed by PAL Robotics, it can perform autonomous navigation in various surroundings and has voice and face recognition capabilities.
2011	In November Honda unveiled its second generation Honda Asimo Robot. The all new Asimo is the first version of the robot with semi-autonomous capabilities.
2012	In April, the Advanced Robotics Department in Italian Institute of Technology released its first version of the COmpliant huMANoid robot COMAN which is designed for robust dynamic walking and balancing in rough terrain.

HEXAPOD (ROBOTICS)

Fig. A desktop sized hexapod.

A hexapod robot is a mechanical vehicle that walks on six legs. Since a robot can be statically stable on three or more legs, a hexapod robot has a great deal of flexibility in how it can move. If legs become disabled, the robot may still be able to walk. Furthermore, not all of the robot's legs are needed for stability; other legs are free to reach new foot placements or manipulate a payload.

Many hexapod robots are biologically inspired by Hexapoda locomotion. Hexapods may be used to test biological theories about insect locomotion, motor control, and neurobiology.

DESIGNS

Fig. Two hexapod robots at the Georgia Institute of Technology with CMUCams mounted on top

Hexapod designs vary in leg arrangement. Insect-inspired robots are typically laterally symmetric, such as the RiSE robot at Carnegie Mellon. A radially symmetric hexapod is ATHLETE (All-Terrain Hex-Legged Extra-Terrestrial Explorer) robot at JPL.

Typically, individual legs range from two to six degrees of freedom. Hexapod feet are typically pointed, but can also be tipped with adhesive material to help climb walls or wheels so the robot can drive quickly when the ground is flat.

LOCOMOTION

Most often, hexapods are controlled by gaits, which allow the robot to move forward, turn, and perhaps side-step. Some of the most common gaits are as follows:

- Alternating tripod: 3 legs on the ground at a time.
- Quadruped.
- Crawl: move just one leg at a time.

Gaits for hexapods are often stable, even in slightly rocky and uneven terrain.

Motion may also be nongaited, which means the sequence of leg motions is not fixed, but rather chosen by the computer in response to the sensed environment. This may be most helpful in very rocky terrain, but existing techniques for motion planning are computationally expensive.

BIOLOGICALLY INSPIRED

Insects are chosen as models because their nervous system is simpler than other animal species. Also, complex behaviours can be attributed to just a few neurons and the pathway between sensory input and motor output is relatively shorter. Insects' walking behaviour and neural architecture are used to improve robot locomotion. Alternatively,biologists can use hexapod robots for testing different hypotheses.

Biologically inspired hexapod robots largely depend on the insect species used as a model. The cockroach and the stick insect are the two most commonly used insect species; both have been ethologically and neurophysiologically extensively studied. At present no complete nervous system is known, therefore, models usually combine different insect models, including those of other insects.

Insect gaits are usually obtained by two approaches: the centralized and the decentralized control architectures. Centralized controllers directly specify transitions of all legs, whereas in decentralized architectures, six nodes (legs) are connected in a parallel network; gaits arise by the interaction between neighbouring legs.

LEG SENSOR INTEGRATION FOR HEXAPOD ROBOT MOTION

This project is a collaboration of many departments on this campus, such as Bioengineering, Electrical and Computer Engineering, Mechanical Engineering, Computer Science, and Neuroscience. This project will be very educational because a lot of professionals are doing research in different fields, collecting the data and eventually transferring it to the Beckman Institute where the physical robot is being built. My main task will be improving the leg motion of the robot by integration of sensors. This sounds like a very exciting project.

Goals and Intentions

The main goals of my project are leg sensor interfacing, software modification to robot controller, sensor data acquisition, and acquiring 'normal' readings from sensors during leg motions, and provide behavioral responses to sensor information. Leg sensor interfacing includes testing of surface mount board, rewiring incorrect traces, and testing new sensors with old sensor algorithm.

Software modifications to robot controller includes sampling sensor data, creating leg motion, and using data to effect leg motion. Sensor data acquisition includes collecting files of data to analyze, looking at data in Matlab, and adjusting the gains on individual sensors. Taking normal readings from sensors during leg motions includes mapping out normal readings during leg motions and looking at situations that create abnormal readings. Providing behavioral reactions to sensor information includes providing reactions such as swinging the leg higher and re-stepping in response to the abnormal readings.

Block Functions

The block diagram consists of sensors that read the data according to the environment and sends to the software which in turn detects the kind of actions that need to take based on the acquired data. The signals from the software get sent to the actuators where the actual physical motion takes place.

There are four types of sensors: foot sensors, strain gages, pressure sensors, and position sensors. The foot sensors detect the ground conditions with the 'Toe' and also detect forces at the 'Toe' to help with the thrusting motion.

The strain gages measure load on a leg segments to balance the weight of the robot. The pressure sensors monitor air pressure in each of the cylinder chambers to measure the force it is generating. The position sensors measure angle between two adjacent leg segments so walking motions can be generated and maintained.

The software reads input from the sensors, decides on leg motion according to the readings, and produces valve signals to move the legs. The software then generates signals, which are sent to the flexor and extensor air valves, to perform the appropriate leg motions.

Performance of the Finished Project

At the end of the project, the requirements for the performance are an adequate amplification of the sensors to be read by the computer input/output (I/O) board and the sensor data must be able to detect different situations the leg will encounter. For example, some sensors have very low output voltages that must be adjusted to a range between 0-5V for the I/O board to be able to read them.

Also, if the leg hits something during a swing, or if it touches the ground, the sensors must be able to detect that these situations and then the software must be able to detect which situation has occurred.

INDUSTRIAL ROBOT

An industrial robot is defined by ISO as an automatically controlled, reprogrammable, multipurpose manipulator programmable in three or more axes. The field of robotics may be more practically defined as the study, design and use of robot systems formanufacturing (a top-level definition relying on the prior definition of robot).

Fig. Articulated industrial robot operating in a foundry.

Typical applications of robots include welding, painting, assembly, pick and place (such as packaging, palletizing and SMT), product inspection, and testing; all accomplished with high endurance, speed, and precision.

ROBOT TYPES, FEATURES

Fig. A set of six-axis robots used forwelding.

Fig. Factory Automation with industrial robots for palletizing food products like bread and toast at a bakery in Germany

The most commonly used robot configurations are articulated robots, SCARA robots, Delta robots and Cartesian coordinate robots, (aka gantry robots or x-y-z robots). In the context of general robotics, most types of robots would fall into the category ofrobotic arms (inherent in the use of the word manipulator in the above-mentioned ISO standard). Robots exhibit varying degrees ofautonomy:

- Some robots are programmed to faithfully carry out specific actions over and over again (repetitive actions) without variation and with a high degree of accuracy. These actions are determined by programmed routines that specify the direction, acceleration, velocity, deceleration, and distance of a series of coordinated motions.
- Other robots are much more flexible as to the orientation of the object on which they are operating or even the task that has to be performed on the object itself, which the robot may even need to identify. For example, for more precise guidance, robots often contain machine vision sub-systems acting as their "eyes", linked to powerful computers or controllers. Artificial intelligence, or what passes for it, is becoming an increasingly important factor in the modern industrial robot.

HISTORY OF INDUSTRIAL ROBOTICS

The earliest known industrial robot, conforming to the ISO definition was completed by "Bill" Griffith P. Taylor in 1937 and published in Meccano Magazine, March 1938. The crane-like device was built almost entirely using Meccano parts, and powered by a single electric motor. Five axes of movement were possible, including Grab and Grab Rotation.

Automation was achieved using punched paper tape to energise solenoids, which would facilitate the movement of the crane's control levers. The robot could stack wooden blocks in pre-programmed patterns. The number of motor revolutions required for each desired movement was first plotted on graph paper. This information was then transferred to the paper tape, which was also driven by the robot's single motor. Chris Shute built a complete replica of the robot in 1997.

George Devol applied for the first robotics patents in 1954 (granted in 1961). The first company to produce a robot was Unimation, founded by Devol and Joseph F. Engelberger in 1956, and was based on Devol's original patents. Unimation robots were also called programmable transfer machines since their main use at first was to transfer objects from one point to another, less than a dozen feet or so apart.

They used hydraulic actuators and were programmed in joint coordinates, i.e. the angles of the various joints were stored during a teaching phase and replayed in operation. They were accurate to within 1/10,000 of an inch (note:

although accuracy is not an appropriate measure for robots, usually evaluated in terms of repeatability - see later). Unimation later licensed their technology to Kawasaki Heavy Industries and GKN, manufacturing Unimates in Japan and England respectively.

For some time Unimation's only competitor was Cincinnati Milacron Inc. of Ohio. This changed radically in the late 1970s when several big Japanese conglomerates began producing similar industrial robots.

In 1969 Victor Scheinman at Stanford University invented the Stanford arm, an all-electric, 6-axis articulated robot designed to permit an arm solution. This allowed it accurately to follow arbitrary paths in space and widened the potential use of the robot to more sophisticated applications such as assembly and welding. Scheinman then designed a second arm for the MIT AI Lab, called the "MIT arm."

Scheinman, after receiving a fellowship from Unimation to develop his designs, sold those designs to Unimation who further developed them with support from General Motors and later marketed it as the Programmable Universal Machine for Assembly (PUMA).

Industrial robotics took off quite quickly in Europe, with both ABB Robotics and KUKA Robotics bringing robots to the market in 1973. ABB Robotics (formerly ASEA) introduced IRB 6, among the world's first commercially available all electric micro-processor controlled robot. The first two IRB 6 robots were sold to Magnusson in Sweden for grinding and polishing pipe bends and were installed in production in January 1974. Also in 1973 KUKA Robotics built its first robot, known as FAMULUS, also one of the first articulated robots to have six electromechanically driven axes.

Interest in robotics increased in the late 1970s and many US companies entered the field, including large firms like General Electric, and General Motors (which formed joint venture FANUC Robotics with FANUC LTD of Japan). U.S. startup companies included Automatix and Adept Technology, Inc. At the height of the robot boom in 1984, Unimation was acquired by Westinghouse Electric Corporation for 107 million U.S. dollars. Westinghouse sold Unimation to Stäubli Faverges SCA of France in 1988, which is still making articulated robots for general industrial and cleanroom applications and even bought the robotic division of Bosch in late 2004.

Only a few non-Japanese companies ultimately managed to survive in this market, the major ones being: Adept Technology, Stäubli-Unimation, the Swedish-Swiss companyABB Asea Brown Boveri, the German company KUKA Robotics and the Italian company Comau.

TECHNICAL DESCRIPTION

Defining parameters

- Number of axes – two axes are required to reach any point in a plane; three axes are required to reach any point in space. To fully control the orientation of the end of the arm (i.e. the wrist) three more axes (yaw, pitch, and roll) are required. Some designs (e.g. the SCARA robot) trade limitations in motion possibilities for cost, speed, and accuracy.

- Degrees of freedom which is usually the same as the number of axes.
- Working envelope – the region of space a robot can reach.
- Kinematics – the actual arrangement of rigid members and joints in the robot, which determines the robot's possible motions. Classes of robot kinematics include articulated, cartesian, parallel and SCARA.
- Carrying capacity or payload – how much weight a robot can lift.
- Speed – how fast the robot can position the end of its arm. This may be defined in terms of the angular or linear speed of each axis or as a compound speed i.e. the speed of the end of the arm when all axes are moving.
- Acceleration - how quickly an axis can accelerate. Since this is a limiting factor a robot may not be able to reach its specified maximum speed for movements over a short distance or a complex path requiring frequent changes of direction.
- Accuracy – how closely a robot can reach a commanded position. When the absolute position of the robot is measured and compared to the commanded position the error is a measure of accuracy. Accuracy can be improved with external sensing for example a vision system or Infra-Red. See robot calibration. Accuracy can vary with speed and position within the working envelope and with payload.
- Repeatability - how well the robot will return to a programmed position. This is not the same as accuracy. It may be that when told to go to a certain X-Y-Z position that it gets only to within 1 mm of that position. This would be its accuracy which may be improved by calibration. But if that position is taught into controller memory and each time it is sent there it returns to within 0.1mm of the taught position then the repeatability will be within 0.1mm.

Accuracy and repeatability are different measures. Repeatability is usually the most important criterion for a robot and is similar to the concept of 'precision' in measurement - see Accuracy and precision. ISO 9283 sets out a method whereby both accuracy and repeatability can be measured. Typically a robot is sent to a taught position a number of times and the error is measured at each return to the position after visiting 4 other positions. Repeatability is then quantified using the standard deviation of those samples in all three dimensions.

A typical robot can, of course make a positional error exceeding that and that could be a problem for the process. Moreover the repeatability is different in different parts of the working envelope and also changes with speed and payload. ISO 9283 specifies that accuracy and repeatability should be measured at maximum speed and at maximum payload.

But this results in pessimistic values whereas the robot could be much more accurate and repeatable at light loads and speeds. Repeatability in an industrial process is also subject to the accuracy of the end effector, for example a gripper, and even to the design of the 'fingers' that match the gripper to the object being grasped.

For example if a robot picks a screw by its head the screw could be at a random angle. A subsequent attempt to insert the screw into a hole could easily fail. These and similar scenarios can be improved with 'lead-ins' e.g. by making the entrance to the hole tapered.

- Motion control – for some applications, such as simple pick-and-place assembly, the robot need merely return repeatably to a limited number of pre-taught positions. For more sophisticated applications, such as welding and finishing (spray painting), motion must be continuously controlled to follow a path in space, with controlled orientation and velocity.
- Power source – some robots use electric motors, others use hydraulic actuators. The former are faster, the latter are stronger and advantageous in applications such as spray painting, where a spark could set off an explosion; however, low internal air-pressurisation of the arm can prevent ingress of flammable vapours as well as other contaminants.
- Drive – some robots connect electric motors to the joints via gears; others connect the motor to the joint directly (direct drive). Using gears results in measurable 'backlash' which is free movement in an axis. Smaller robot arms frequently employ high speed, low torque DC motors, which generally require high gearing ratios; this has the disadvantage of backlash. In such cases the harmonic drive is often used.
- Compliance - this is a measure of the amount in angle or distance that a robot axis will move when a force is applied to it. Because of compliance when a robot goes to a position carrying its maximum payload it will be at a position slightly lower than when it is carrying no payload. Compliance can also be responsible for overshoot when carrying high payloads in which case acceleration would need to be reduced.

Robot programming and interfaces

The setup or programming of motions and sequences for an industrial robot is typically taught by linking the robot controller to alaptop, desktop computer or (internal or Internet) network. A robot and a collection of machines or peripherals is referred to as a workcell, or cell. A typical cell might contain a parts feeder, amolding machine and a robot. The various machines are 'integrated' and controlled by a single computer or PLC. How the robot interacts with other machines in the cell must be programmed, both with regard to their positions in the cell and synchronizing with them.

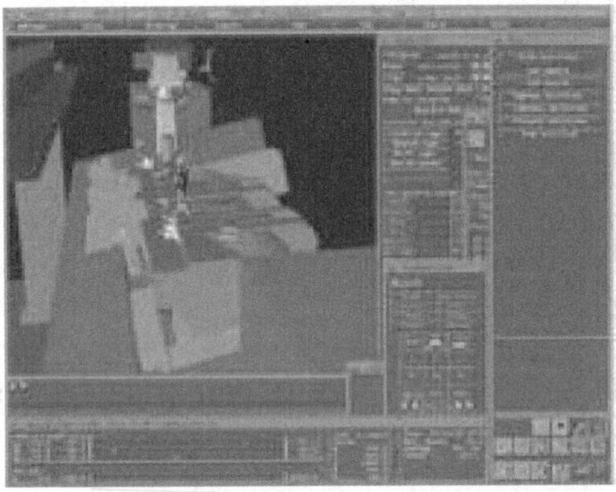

Fig. Offline programming by ROBCAD

Fig. A typical well-used teach pendant with optional mouse

Software: The computer is installed with corresponding interface software. The use of a computer greatly simplifies the programming process. Specialized robot software is run either in the robot controller or in the computer or both depending on the system design.

There are two basic entities that need to be taught (or programmed): positional data and procedure. For example in a task to move a screw from a feeder to a hole the positions of the feeder and the hole must first be taught or programmed. Secondly the procedure to get the screw from the feeder to the hole must be programmed along with any I/O involved, for example a signal to indicate when the screw is in the feeder ready to be picked up. The purpose of the robot software is to facilitate both these programming tasks. Teaching the robot positions may be achieved a number of ways:

Positional commands The robot can be directed to the required position using a GUI or text based commands in which the required X-Y-Z position may be specified and edited.

Teach pendant: Robot positions can be taught via a teach pendant. This is a handheld control and programming unit. The common features of such units are the ability to manually send the robot to a desired position, or "inch" or "jog" to adjust a position. They also have a means to change the speed since a low speed is usually required for careful positioning, or while test-running through a new or modified routine. A large emergency stop button is usually included as well. Typically once the robot has been programmed there is no more use for the teach pendant.

Lead-by-the-nose is a technique offered by many robot manufacturers. In this method, one user holds the robot's manipulator, while another person enters a command which de-energizes the robot causing it to go limp. The user then moves the robot by hand to the required positions and/or along a required path while the software logs these positions into memory. The program can later run the robot to these positions or along the taught path. This technique is popular for tasks such as paint spraying.

Offline programming is where the entire cell, the robot and all the machines or instruments in the workspace are mapped graphically. The robot can then be moved on screen and the process simulated. The technique has limited value because it relies on accurate measurement of the positions of the associated equipment and also relies on the positional accuracy the robot which may or may not conform to what is programmed.

Others In addition, machine operators often use user interface devices, typically touchscreen units, which serve as the operator control panel. The operator can switch from program to program, make adjustments within a program and also operate a host of peripheral devices that may be integrated within the same robotic system. These includeend effectors, feeders that supply components to the robot, conveyor belts, emergency stop controls, machine vision systems, safety interlock systems, bar code printers and an almost infinite array of other industrial devices which are accessed and controlled via the operator control panel.

The teach pendant or PC is usually disconnected after programming and the robot then runs on the program that has been installed in its controller. However a computer is often used to 'supervise' the robot and any peripherals, or to provide additional storage for access to numerous complex paths and routines.

End-of-arm Tooling

The most essential robot peripheral is the end effector, or end-of-arm-tooling (EOT). Common examples of end effectors include welding devices (such as MIG-welding guns, spot-welders, etc.), spray guns and also grinding and deburring devices (such as pneumatic disk or belt grinders, burrs, etc.), and grippers (devices that can grasp an object, usually electromechanical or pneumatic). Another common means of picking up an object is by vacuum. End effectors are frequently highly complex, made to match the handled product and often capable of picking up an array of products at one time. They may utilize various sensors to aid the robot system in locating, handling, and positioning products.

Controlling Movement

For a given robot the only parameters necessary to completely locate the end effector (gripper, welding torch, etc.) of the robot are the angles of each of the joints or displacements of the linear axes (or combinations of the two for robot formats such as SCARA). However there are many different ways to define the points.

The most common and most convenient way of defining a point is to specify a Cartesian coordinate for it, i.e. the position of the 'end effector' in mm in the X, Y and Z directions relative to the robot's origin. In addition, depending on the types of joints a particular robot may have, the orientation of the end effector in yaw, pitch, and roll and the location of the tool point relative to the robot's faceplate must also be specified. For a jointed arm these coordinates must be converted to joint angles by the robot controller and such conversions are known as Cartesian Transformations which may need to be performed iteratively or recursively for a multiple axis robot. The mathematics of the relationship between joint angles and actual spatial coordinates is called kinematics.

Positioning by Cartesian coordinates may be done by entering the coordinates into the system or by using a teach pendant which moves the robot in X-Y-Z directions. It is much easier for a human operator to visualize motions up/down, left/right, etc. than to move each joint one at a time. When the desired position is reached it is then defined in some way particular to the robot software in use, e.g. P1 - P5 below.

Typical Programming

Most articulated robots perform by storing a series of positions in memory, and moving to them at various times in their programming sequence. For example, a robot which is moving items from one place to another might have a simple 'pick and place' program similar to the following:

Define points P1–P5:

1. Safely above workpiece (defined as P1)
2. 10 cm Above bin A (defined as P2)
3. At position to take part from bin A (defined as P3)

4. 10 cm Above bin B (defined as P4)
5. At position to take part from bin B. (defined as P5)

Define program:

1. Move to P1
2. Move to P2
3. Move to P3
4. Close gripper
5. Move to P2
6. Move to P4
7. Move to P5
8. Open gripper
9. Move to P4
10. Move to P1 and finish

For examples of how this would look in popular robot languages see industrial robot programming.

Singularities

The American National Standard for Industrial Robots and Robot Systems — Safety Requirements (ANSI/RIA R15.06-1999) defines a singularity as "a condition caused by the collinear alignment of two or more robot axes resulting in unpredictable robot motion and velocities." It is most common in robot arms that utilize a "triple-roll wrist". This is a wrist about which the three axes of the wrist, controlling yaw, pitch, and roll, all pass through a common point.

An example of a wrist singularity is when the path through which the robot is traveling causes the first and third axes of the robot's wrist (i.e. robot's axes 4 and 6) to line up. The second wrist axis then attempts to spin 360° in zero time to maintain the orientation of the end effector. Another common term for this singularity is a "wrist flip". The result of a singularity can be quite dramatic and can have adverse effects on the robot arm, the end effector, and the process. Some industrial robot manufacturers have attempted to side-step the situation by slightly altering the robot's path to prevent this condition. Another method is to slow the robot's travel speed, thus reducing the speed required for the wrist to make the transition. The ANSI/RIA has mandated that robot manufacturers shall make the user aware of singularities if they occur while the system is being manually manipulated.

A second type of singularity in wrist-partitioned vertically articulated six-axis robots occurs when the wrist center lies on a cylinder that is centered about axis 1 and with radius equal to the distance between axes 1 and 4. This is called a shoulder singularity. Some robot manufacturers also mention alignment singularities, where axes 1 and 6 become coincident. This is simply a sub-case of shoulder singularities. When the robot passes close to a shoulder singularity, joint 1 spins very fast.

The third and last type of singularity in wrist-partitioned vertically articulated six-axis robots occurs when the wrist's center lies in the same plane as axes 2 and 3.

Singularities are closely related to the phenomena of Gimbal Lock, which has a similar root cause of axes becoming lined up.

A video illustrating these three types of singular configurations is available here.

RECENT AND FUTURE DEVELOPMENTS

As of 2005, the robotic arm business is approaching a mature state, where they can provide enough speed, accuracy and ease of use for most of the applications. Vision guidance (aka machine vision) is bringing a lot of flexibility to robotic cells. However, the end effector attached to a robot is often a simple pneumatic, 2-position chuck. This does not allow the robotic cell to easily handle different parts, in different orientations.

Hand-in-hand with increasing off-line programmed applications, robot calibration is becoming more and more important in order to guarantee a good positioning accuracy.

Other developments include downsizing industrial arms for light industrial use such as production of small products, sealing and dispensing, quality control, handling samples in the laboratory. Such robots are usually classified as "bench top" robots. Robots are used in pharmaceutical research in a technique called High-throughput screening. Bench top robots are also used in consumer applications (micro-robotic arms). Industrial arms may be used in combination with or even mounted on automated guided vehicles (AGVs) to make the automation chain more flexible between pick-up and drop-off.

STEREO VISION FOR MOBILE ROBOTICS

Two eyes or cameras looking at the same scene from different perspectives provide a mean for determining three-dimensional shape and position. Scientific investigation of this effect (called variously stereo vision, stereopsis or single vision) has a rich history in psychology, biology and more recently, in the computational model of perception. Stereo is an important method for machine perception because it leads to direct depth measurements. Additionally, unlike monocular techniques, stereo does not infer depth from weak or unverifiable photometric and statistical assumptions, nor does it require specific detailed objects models. Once stereo images have been brought into point-to-point correspondence, recovering depth by triangulation is straightforward.

Current range imagers have achieved real-time or near real-time performance on images of modest size. For example, stereo algorithms on standard hardware are capable of returning dense 128 x 128 range images at

10 Hz, while scanning laser range-finders can operate at 2 Hz on 256 x 256 images. To take advantage of these devices, researchers have proposed numerous methods for extracting 3-D information from range images. These methods operate either in 3-D Cartesian space (volumetric representations) or in a 2.5-D range image space (contour map method). Contour map methods are particularly attractive for computation bound applications such as mobile robots.

We begin with a discussion of the geometric basis and a computational model for stereo vision. Next, we briefly describe biological aspects of depth perception and a contour map method for depth processing. Finally, we present an obstacle avoidance technique for mobile robots using real-time stereo vision.

STEREO VISION

The geometric basis key problem in stereo vision is to find corresponding points in stereo images. Corresponding points are the projections of a single 3D point in the different image spaces. The difference in the position of corresponding points in their respective images is called disparity. Disparity is a function of both the position of the 3D scene point and of the position, orientation, and physical characteristics of the stereo devices (e.g. cameras).

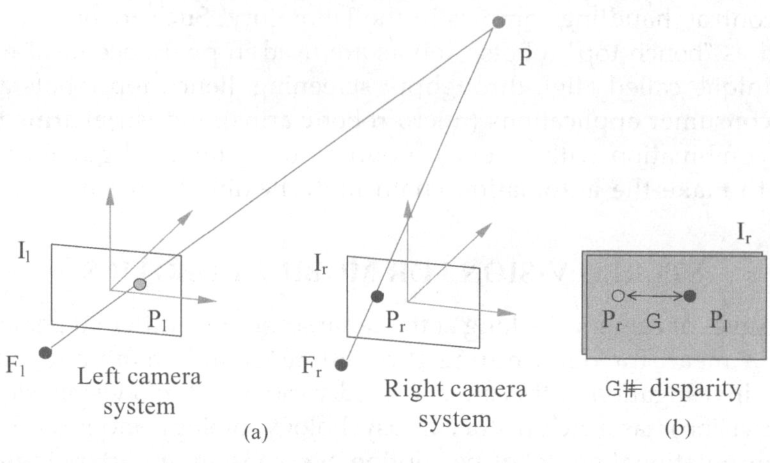

Fig. (a) A system with two cameras is shown. The focal points are Fl and Fr, the image planes are Il and Ir. A point P in the 3D scene is projected onto Pl in the left image and onto Pr in the right image, (b) cyclopean view: the disparity ?? is the difference in the position of the projection of the point P onto the two stereo image planes.

In addition to providing the function that maps pair of corresponding images points onto scene points, a camera model can be used to constraint the search for corresponding image point to one dimension. Any point in the 3D world space together with the centers of projection of two cameras systems, defines an epipolar plane. The intersection of such a plane with an image plane is called an epipolar line. Every point of a given epipolar line must correspond to a single point on the corresponding epipolar line. The search for a match of a point in the first image may therefore be reduced to a one-dimensional neighborhood in the second image plane (as opposed to a 2D neighborhood).

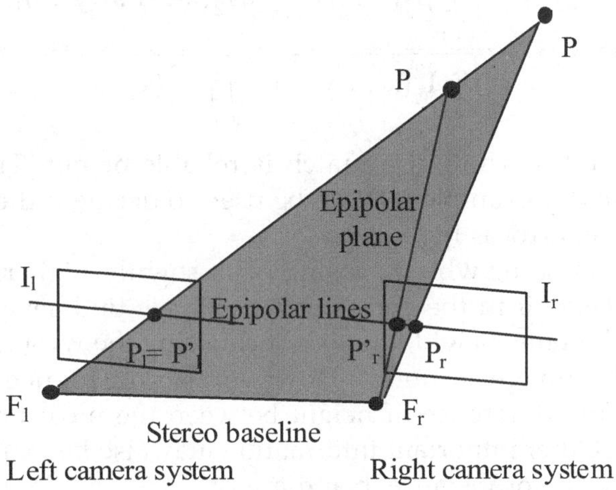

Fig. Epipolar lines and epipolar planes.

When the stereo cameras are oriented such that there is a known horizontal displacement between them, disparity can only occur in the horizontal direction and the stereo images are said to be in correspondence. When a stereo pair is in correspondence, the epipolar lines are coincident with the horizontal scan lines of the digitized pictures.

Ideally, one would like to find the correspondence of every individual pixel in both images of a stereo pair. However, it is obvious that the information content in the intensity value of a single pixel is too low for unambiguous matching. In practice, continuous areas of image intensity are the basic units that are matched. This approach (called area matching) usually involves some form of cross-correlation to establish correspondences.

Matching

The main problem in matching is to find an effective definition of what we call a valid correlation.

Correlation scores are computed by comparing a fixed window in the first image against a shifting window in the second. The second window is moved in the second image by integer increments along the corresponding epipolar line and a correlation score curve is generated for integer disparity values. The measured disparity can then be taken to be the one that provides the largest peak. To quantify the similarity between two correlation windows, we must choose among many different criteria that produce reliable results in a minimum computation time. We denote by I1(x,y) and I2(x,y) the intensity value at pixel (x,y). The correlation window has dimensions (2n . 1) „ (2m . 1).

Therefore, the indexes which appear in the formula below vary between -n and +n for the i-index and between -m and +m for the j-index:

$$C_1(x,y,G) @ \frac{\hat{1}_{i,j} [I_1(x . i, y . j) \circ I_2(x . G . i, y . j)]^2}{\sqrt{\hat{1}_{i,j} I_1^2(x . i, y . j)} „ \sqrt{\hat{1}_{i,j} I_2^2(x . G . i, y . j)}}$$

It is important to know if a match is reliable or not. The form of the correlation curve (for example C1) can be used to decide if the probability of the match to be an error is high or not.

Indeed, errors occur when a wrong peak slightly higher than the right one is chosen. Thus, if in the correlation curve we find several peaks with approximately the same height, the risk of choosing the wrong one increases, especially if the image is noisy. However, a confidence coefficient ?, proportional to the difference of height between the most important peaks may be defined. Other important information may also be extracted from the correlation curve as, for instance, bland areas.

Human depth perception

For human beings, correlation (as described in the previous section) is only a local mechanism of stereoscopic vision [Bruce and Green 1990]. However, imagine the following experiment:

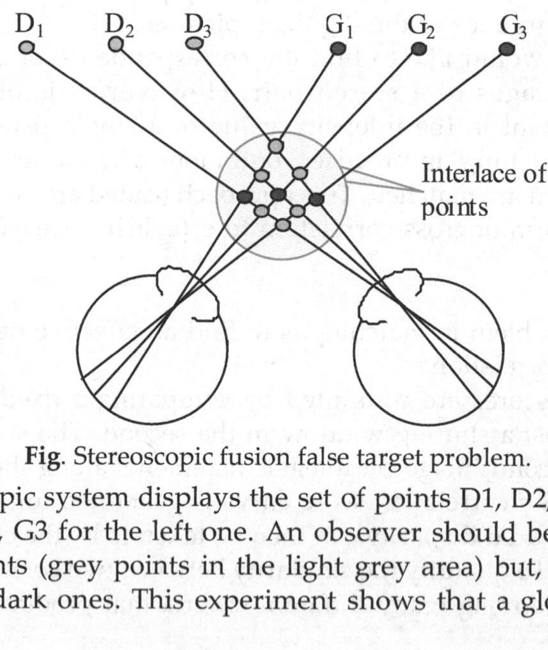

Fig. Stereoscopic fusion false target problem.

A stereoscopic system displays the set of points D1, D2, D3 for the right eye and G1, G2, G3 for the left one. An observer should be able to see any interlace of points (grey points in the light grey area) but, instead, they all succeed to the dark ones. This experiment shows that a global mechanism, based on criteria other than local correlation, is used. Among these ones, the following are taken in account:

- A principle of correlation based on the contours of the image;
- A mechanism of cognitive interpretation which has, in some cases, more priority than the local mechanism of stereo vision [Maar 1982];
- A mechanism of pictorial clues of depth (relative size, relative height, perspective, shade, "fog" effect and interposition;
- A principle of dynamic clues, such as motion parallax;
- Other mechanisms such as correlation of frequency filtered images [Poggio and Poggio 1984];

(a) (b)

Fig. (a) "Illusory" contours defining a square giving the impression that this shape is placed in front of 4 circles, (b) interposition principle (cognitive interpretation).

This list is not exhaustive but presents the most significant criteria belonging to the global system of the human stereoscopic vision.

In summary, the human stereo system uses a number of interesting methods, which work together to recover depth. On one hand, this system is very powerful because even using one eye, it is possible to perceive depth. On the other hand, it is also very subjective because a trompe-l'œil can fool our perceptive system.

In the next section, we will focus on an example of computational model for range (disparity) image parsing.

Contour map method

For many applications, working in 3D Euclidean space turns out to be unnecessary and difficult to manage. To reduce the amount of data which has to be processed, we introduce a method of quantifying volumes that allows us to manipulate range images directly, without having to first transform to 3-D space.

The method is similar to the use of contour maps to represent elevations; hence, we call it the contour method [Chauvin, Marti and Konolige 1997]. A contour represents the elevation at a particular height; all terrain between one contour line and the next is at an elevation between that represented by the contour lines. Contour creation can be visualized as a set of planes parallel to the ground at specified heights, intersecting the terrain. These cutting planes induce a quantization of the 3-D space based on elevation. Our approach uses this basic idea, and consists of the following steps:

1. Constructing a set of volumes in 3-D space using a set of cutting surfaces (not necessarily planar);
2. Projecting the cutting surfaces back to the range image to induce a quantization of the range data;
3. Using the quantized range image to construct terrain models or other abstractions;

In cases where the desired segmentation is relative to the sensor viewpoint, the first two steps can be achieved off-line, leading to significant computational savings, especially when the cutting surfaces are complicated. In addition, step 3 can often be performed in the range image space, which is much more efficient than working in the volumetric space. Also, in contrast to the grid-based approaches, the cutting surfaces need not be regular, and can be sized to take in account the precision and error characteristics of the range data.

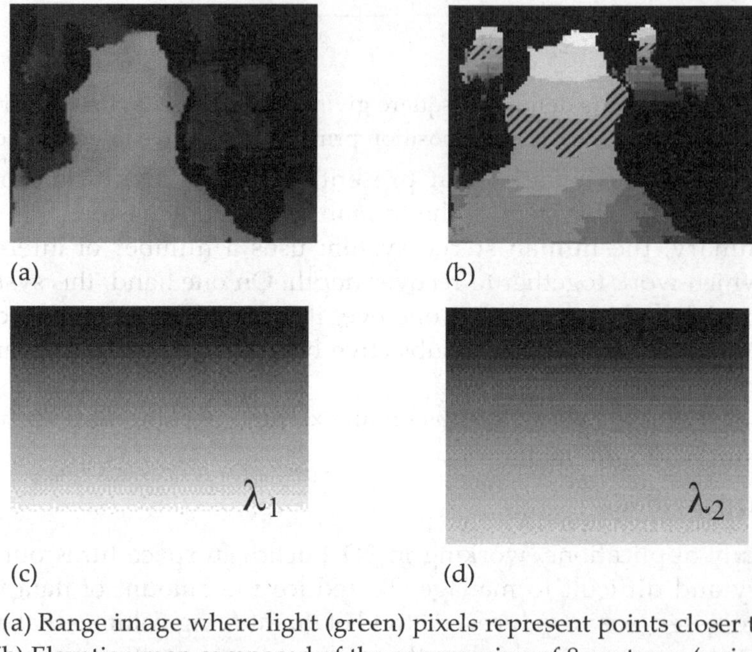

Fig. (a) Range image where light (green) pixels represent points closer than dark ones, (b) Elevation map composed of the superposing of 8 contours (a single contour is represented with a special pattern), (c) and (d) projection of the cutting surfaces back in the range image for the special contour of image b.

MOBILE ROBOT OBSTACLE AVOIDANCE

The contour method is well suited for use with the vector field histogram

(VFH) algorithm for mobile robot obstacle avoidance [Borenstein and Koren 1991]. Originally developed with sonar sensors, the method used three steps:

1. A regular 2-D histogram grid in plan view, holding the results of sonar sensor readings around the robot. The value of each grid point represents the number of sonar readings that indicated an object within the point;
2. A polar histogram is computed from the histogram grid, with k regular angular sectors instead of a rectilinear grid. The value hk of each sector in the polar histogram represents the obstacle density in that direction;
3. Steering and velocity values are extracted from the polar histogram;

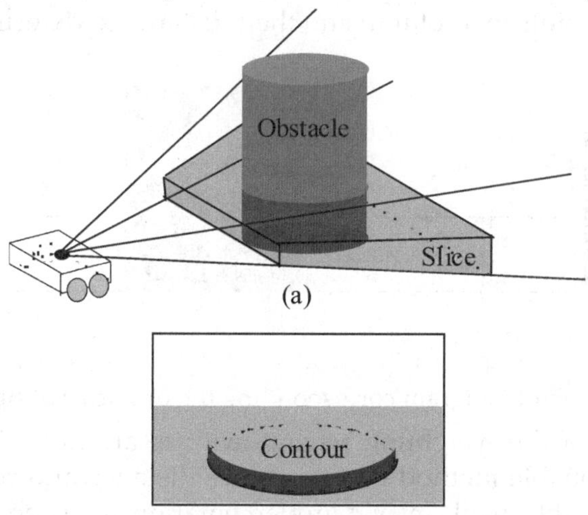

Fig. Representation of an obstacle (a) in the Cartesian space, (b) in the contour image space.

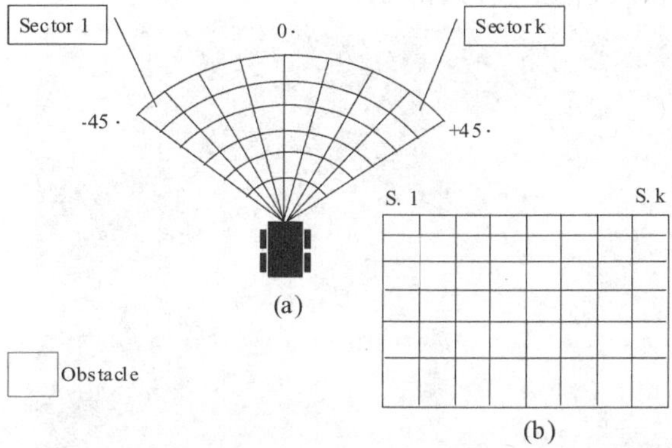

Fig. Two obstacles (a) in the polar grid, (b) in the contour image grid.

In range images, each column of the image represents a polar sector whose angular width is determined by the camera parameters. We let the k sectors correspond to the columns of the range image. Thus, we can construct the polar histogram directly from the contour representation, without having to convert to Cartesian space.

The key step is calculating the histogram value hk for each sector. Roughly speaking, this value represents the probability of finding an obstacle close to the robot in the direction of sector k. The simplest idea is to use a single cutting surface at elevation over the ground plane sufficient to constitute an obstacle for the robot.

Any points in the resulting contour are obstacles, and we can use the number of such points in a column and their distance to determine a histogram value.

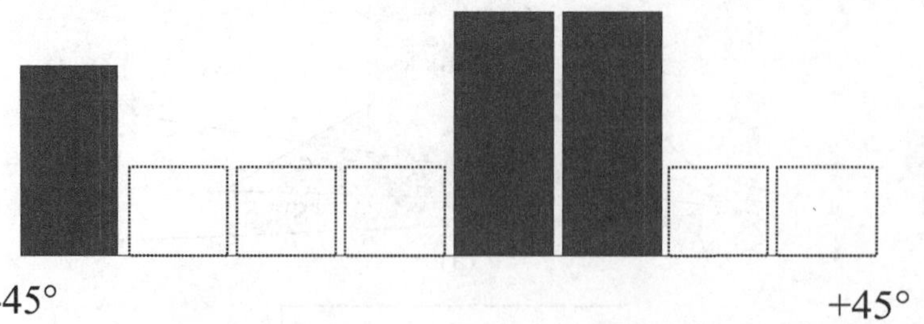

Fig. Polar histogram corresponding the obstacles of figure.

The details of the weighting scheme we use are not critical; we expect almost any reasonable method that combines distance and number of points will work reasonably well. In our implementation we used a stereo system with disparity as the range metric, and let each contour cell m contributes In [r(m)] to its sector value.

This measure compensates for the fact that the disparity increases hyperbolically as an object gets closer.

(a)

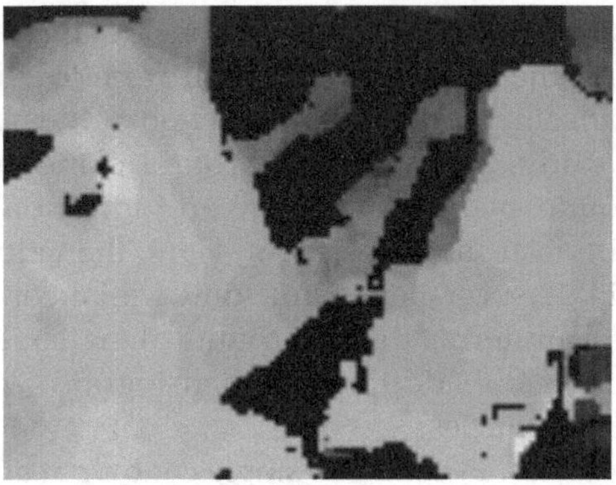

(b)

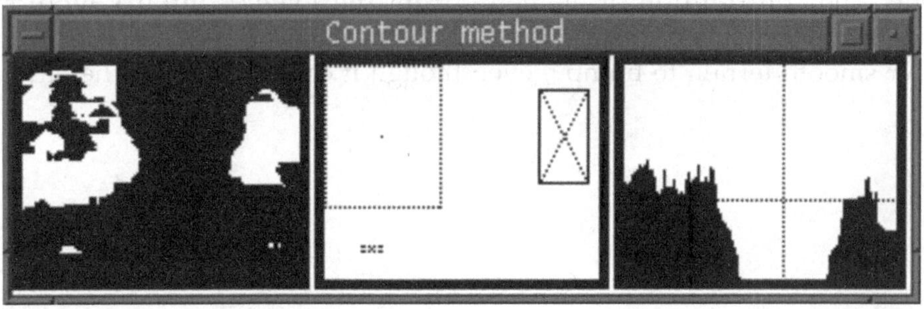

(c)

Fig. Experimental results. (a) image of the scene, (b) corresponding disparity image, (c) from left to right: contour, object detection and polar histogram (the vertical line corresponds to the direction followed by the robot and the horizontal line, its speed).

The VFH method was implemented using a small stereo system for range images and a PC for processing the VFH algorithm. The stereo system returned images at a 5 Hz rate, and the VFH processing took less than 10 ms per image to format the polar histogram and extract the desired direction and speed of travel. Data were then sent to a robot navigation program in order to steer a Koala or Pioneer robot.

Figure shows a calculation of the polar histogram from a typical range image and a single-contour segmentation. In this case, the sensor covers about a 70 degree angle, and each sector is about 0.5 degrees. Image (a) is an intensity image of the scene, and (b) shows the disparity map computed by a stereo system.

Brighter green values are higher disparities, hence closer to the camera. The final set of images (c) shows the contour (left side), a segmentation of some obstacles (middle), and finally the polar histogram. The middle of the histogram is straight along the camera optical axis, and the vertical line indicates the direction of travel that the VFH algorithm has found. From the picture, this is the direction through the open door.

Because of the small sector size, there is considerable variation in adjacent histogram values, and the result could benefit from low-pass filtering.

Additional enhancements in the construction of the polar histogram from contours are under study, among them: Ground plane detection. A contour representing the ground plane would give an indication that there actually was a reasonable path in front of the robot. Here, the value In [r(m)] for each cell in a sector would be subtracted from some initial constant for the sector;

Holes. A contour underneath the ground plane could be used to check for holes near the robot. The addition to the histogram would be the same as for positive-elevation obstacles;

Small-height obstacles. Instead of a single contour at an appropriate height for obstacles, several could be positioned starting from just over the ground plane. The contribution of elements from the lower contours would be weighted by a fraction q depending on their height. Thus, the robot would prefer smooth terrain to bumpy, even though it could negotiate the latter;

2

Autonomous Robot

INTRODUCTION

Autonomous robots are robots that can perform with a high degree of autonomy, which is particularly desirable in fields such as space exploration, cleaning floors, mowing lawns, and waste water treatment.

Some modern factory robots are "autonomous" within the strict confines of their direct environment. It may not be that every degree of freedom exists in their surrounding environment, but the factory robot's workplace is challenging and can often contain chaotic, unpredicted variables. The exact orientation and position of the next object of work and (in the more advanced factories) even the type of object and the required task must be determined. This can vary unpredictable (at least from the robot's point of view).

One important area of robotics research is to enable the robot to cope with its environment whether this be on land, underwater, in the air, underground, or in space.

A fully autonomous robot has the ability to

- Gain information about the environment (Rule #1)
- Work for an extended period without human intervention (Rule #2)
- Move either all or part of itself throughout its operating environment without human assistance (Rule #3)
- Avoid situations that are harmful to people, property, or itself unless those are part of its design specifications (Rule #4)

An autonomous robot may also learn or gain new knowledge like adjusting for new methods of accomplishing its task(s) or adapting to changing surroundings. Autonomous robots still require regular maintenance, as do other machines.

EXAMPLES OF PROGRESS TOWARDS COMMERCIAL AUTONOMOUS ROBOTS

Self-maintenance

The first requirement for complete physical autonomy is the ability for a robot to take care of itself. Many of the battery powered robots on the market today can find and connect to a charging station, and some toys like Sony's Aibo are capable of self-docking to charge their batteries.

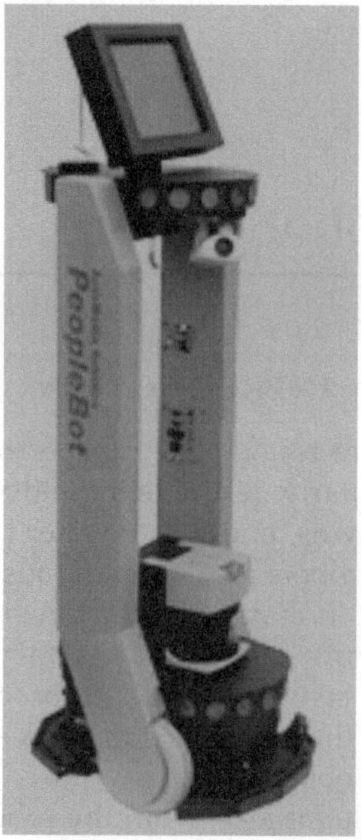

Fig. Exteroceptive sensors: 1. blue laser rangefinder senses up to 360 distance readings in a 180-degree slice; 2. 24 round golden ultrasonic sensors sample range readings in a 15-degree cone; 3. ten touch panels along the bottom detect shoes and other low-lying objects. 4. break beams between the lower and upper segments sense tables and other mid-level obstacles.

Self maintenance is based on "proprioception", or sensing one's own internal status. In the battery charging example, the robot can tell proprioceptively that its batteries are low and it then seeks the charger. Another common proprioceptive sensor is for heat monitoring. Increased proprioception will be required for robots to work autonomously near people and in harsh environments.

Common proprioceptive sensors are:

- Thermal
- Hall Effect
- Optical
- Contact.

Sensing the environment

Exteroception is sensing things about the environment. Autonomous robots must have a range of environmental sensors to perform their task and stay out of trouble.

Common exteroceptive sensors are:

- Electromagnetic spectrum
- Sound
- Touch
- Chemical sensors (smell, odor)
- Temperature
- Range to things in the environment
- Attitude (Inclination)

Some robotic lawn mowers will adapt their programming by detecting the speed in which grass grows as needed to maintain a perfect cut lawn, and some vacuum cleaning robots have dirt detectors that sense how much dirt is being picked up and use this information to tell them to stay in one area longer.

Task performance

The next step in autonomous behavior is to actually perform a physical task. A new area showing commercial promise is domestic robots, with a flood of small vacuuming robots beginning with iRobot and Electrolux in 2002. While the level of intelligence is not high in these systems, they navigate over wide areas and pilot in tight situations around homes using contact and non-contact sensors. Both of these robots use proprietary algorithms to increase coverage over simple random bounce.

The next level of autonomous task performance requires a robot to perform conditional tasks. For instance, security robots can be programmed to detect intruders and respond in a particular way depending upon where the intruder is.

Indoor position sensing and navigation

For a robot to associate behaviors with a place (localization) requires it to know where it is and to be able to navigate point-to-point. Such navigation began with wire-guidance in the 1970s and progressed in the early 2000s to beacon-based triangulation. Current commercial robots autonomously navigate based on sensing natural features. The first commercial robots to achieve this were Pyxus' HelpMate hospital robot and the CyberMotion guard robot, both designed by robotics pioneers in the 1980s. These robots originally used manually created CAD floor plans, sonar sensing and wall-following variations to navigate buildings.

The next generation, such as MobileRobots' PatrolBot andautonomous wheelchair, both introduced in 2004, have the ability to create their own laser-based maps of a building and to navigate open areas as well as corridors. Their control system changes its path on the fly if something blocks the way.

At first, autonomous navigation was based on planar sensors, such as laser range-finders, that can only sense at one level. The most advanced systems now fuse information from various sensors for both localization (position) and navigation. Systems such as Motivity can rely on different sensors in different areas, depending upon which provides the most reliable data at the time, and can re-map a building autonomously. Rather than climb stairs, which requires highly specialized hardware, most indoor robots navigate handicapped-accessible areas, controlling elevators and electronic doors.With such electronic access-control interfaces, robots can now freely navigate indoors. Autonomously climbing stairs and opening doors manually are topics of research at the current time. As these indoor techniques continue to develop, vacuuming robots will gain the ability to clean a specific user-specified room or a whole floor. Security robots will be able to cooperatively surround intruders and cut off exits. These advances also bring concomitant protections: robots' internal maps typically permit "forbidden areas" to be defined to prevent robots from autonomously entering certain regions.

Outdoor autonomous position-sensing and navigation

Outdoor autonomy is most easily achieved in the air, since obstacles are rare. Cruise missiles are rather dangerous highly autonomous robots. Pilotless drone aircraft are increasingly used for reconnaissance. Some of these unmanned aerial vehicles (UAVs) are capable of flying their entire mission without any human interaction at all except possibly for the landing where a person intervenes using radio remote control. But some drone aircraft are capable of a safe, automatic landing also.

Fig. The Seekur and MDARS robots demonstrate their autonomous navigation and security capabilities at an airbase.

Outdoor autonomy is the most difficult for ground vehicles, due to: a) 3-dimensional terrain; b) great disparities in surface density; c) weather exigencies and d) instability of the sensed environment.

In the US, the MDARS project, which defined and built a prototype outdoor surveillance robot in the 1990s, is now moving into production and will be implemented in 2006. The General Dynamics MDARS robot can navigate semi-autonomously and detect intruders, using the MRHA software architecture planned for all unmanned military vehicles. The Seekur robot was the first commercially available robot to demonstrate MDARS-like capabilities for general use by airports, utility plants, corrections facilities and Homeland Security.

The Mars rovers MER-A and MER-B (now known as Spirit rover and Opportunity rover) can find the position of the sun and navigate their own routes to destinations on the fly by:

- Mapping the surface with 3-D vision
- Computing safe and unsafe areas on the surface within that field of vision
- Computing optimal paths across the safe area towards the desired destination
- Driving along the calculated route;
- Repeating this cycle until either the destination is reached, or there is no known path to the destination

The planned ESA Rover, ExoMars Rover, is capable of vision based relative localisation and absolute localisation to autonomously navigate safe and efficient trajectorys to targets by:

- Reconstructing 3D models of the terrain surrounding the Rover using a pair of stereo cameras
- Determining safe and unsafe areas of the terrain and the general 'difficulty' for the Rover to navigate the terrain
- Computing efficient paths across the safe area towards the desired destination
- Driving the Rover along the planned path
- Building up a 'Navigation Map' of all past navigation data

The DARPA Grand Challenge and DARPA Urban Challenge have encouraged development of even more autonomous capabilities for ground vehicles, while this has been the demonstrated goal for aerial robots since 1990 as part of the AUVSI International Aerial Robotics Competition.

OPEN PROBLEMS IN AUTONOMOUS ROBOTICS

There are several open problems in autonomous robotics which are special to the field rather than being a part of the general pursuit of AI. According to George A. Bekey's Autonomous Robots: From Biological Inspiration to

Implementation and Control, problems include things such as making sure the robot is able to function correctly and not run into obstacles autonomously.

Energy autonomy & foraging

Researchers concerned with creating true artificial life are concerned not only with intelligent control, but further with the capacity of the robot to find its own resources through foraging (looking for food, which includes both energy and spare parts). This is related to autonomous foraging, a concern within the sciences of behavioral ecology, social anthropology, and human behavioral ecology; as well as robotics,artificial intelligence, and artificial life.

AUTONOMOUS MOBILE ROBOT AND HARDWARE

Welcome to "Robot Boot Camp." In this two-part article series, I'll explain what you can do with a basic mobile machine, a few sensors, and behavioral programming techniques. Behavioral programming provides distinct advantages over other programming techniques. It is independent of any environmental model, and it is more robust in the face of sensor error, and the behaviors can be stacked and run concurrently.

My objectives for my recent robot design were fairly modest. I wanted to build a robot that could cruise on its own, avoid obstacles, escape from inadvertent collisions, and track a light source. I knew that if I could meet such objective other more complex behaviors would be possible (e.g., self-docking on low power).

There certainly many commercial robots on the market that could have met my requirements. But I decided that my best bet would be to roll my own. I wanted to keep things simple, and I wanted to fully understand the sensors and controls for behavioral autonomous operation. The TOMBOT is the fruit of that labor. A colleague came up with the name TOMBOT in honor of its inventor, and the name kind of stuck.

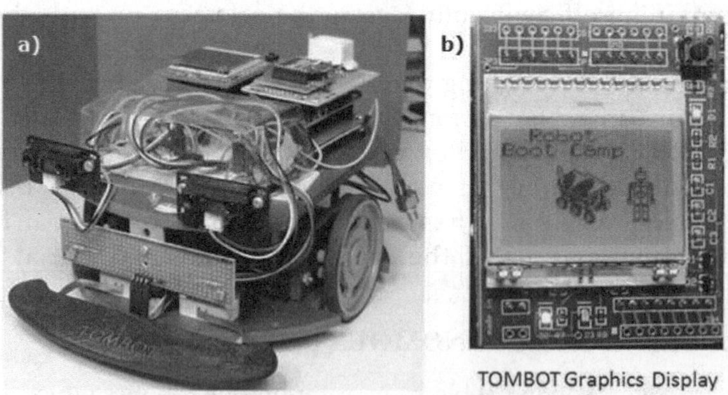

TOMBOT

TOMBOT Graphics Display Content on Power Up

Fig. a—The complete TOMBOT design. b—The graphics display is nice feature.

In this series of articles, I'll present lessons learned and describe the hardware/software design process. The series will detail TOMBOT-style robot hardware and assembly, as well as behavior programming techniques using C code. By the end of the series, I'll have covered a complete behavior programming library and API, which will be available for experimentation.

DESIGN BASICS

The TOMBOT robot is certainly minimal, no frills: two continuous-rotation, variable-speed control servos; two IR (850 nm) analog distance measurement sensors (4- to 30-cm range); two CdS photoconductive cells with good lux response in visible spectrum; and, finally, a front bumper (switch-activated) for collision detection. The platform is simple: servos and sensors on the left and right side of two level platforms. The bottom platform houses bumper, batteries, and servos. The top platform houses sensors and microcontroller electronics. The back part of the bottom platform uses a central skid for balance between the two servos.

Given my background as a Microchip Developer and Academic Partner, I used a Microchip Technology PIC32 microcontroller, a PICkit 3 programmer/debugger, and a free Microchip IDE and 32-bit complier for TOMBOT. (Refer to the TOMBOT components list at the end of this article.)

It was a real thrill to design and build a minimal capability robot that can—with stacking programming behaviors—emulate some "intelligence." TOMBOT is still a work in progress, but I recently had the privilege of demoing it to a first grade class in El Segundo, CA, as part of a Science Technology Engineering and Mathematics (STEM) initiative. The results were very rewarding, but more on that later.

BEHAVIORAL PROGRAMMING

A control system for a completely autonomous mobile robot must perform many complex information-processing tasks in real time, even for simple applications. The traditional method to building control systems for such robots is to separate the problem into a series of sequential functional components. An alternative approach is to use behavioral programming. The technique was introduced by Rodney Brooks out of the MIT Robotics Lab, and it has been very successful in the implementation of a lot of commercial robots, such as the popular Roomba vacuuming. It was even adopted for space applications like NASA's Mars Rover and military seekers.

Programming a robot according to behavior-based principles makes the program inherently parallel, enabling the robot to attend simultaneously to all hazards it may encounter as well as any serendipitous opportunities that may arise. Each behavior functions independently through sensor registration, perception, and action. In the end, all behavior requests are prioritized and arbitrated before action is taken. By stacking the appropriate behaviors, using arbitrated software techniques, the robot appears to show (broadly speaking) "increasing intelligence." The TOMBOT modestly achieves this objective using selective compile configurations to emulate a series of robot behaviors (i.e., Cruise, Home, Escape, Avoid, and Low Power). Figure is a simple model illustration of a behavior program.

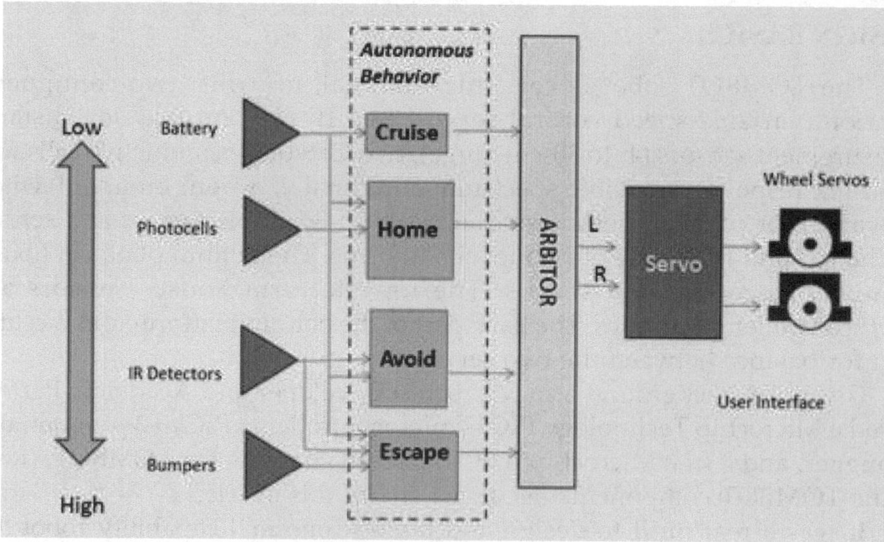

Fig. Behavior program

Joseph Jones's Robot Programming: A Practical Guide to Behavior-Based Robotics (TAB Electronics, 2003) is a great reference book that helped guide me in this effort. It turns out that Jones was part of the design team for the Roomba product.

Debugging a mobile platform that is executing a series of concurrent behaviors can be daunting task. So, to make things easier, I implemented a complete remote control using a wireless link between the robot and a PC. With this link, I can enable or disable autonomous behavior, retrieve the robot sensor status and mode of operations, and curtail and avoid potential robot hazard.

In addition to this, I implemented some additional operator feedback using a small graphics display, LEDs, and a simple sound buzzer. Note the TOMBOT's power-up display in Photo 1b. We take Robot Boot Campvery seriously.

Minimalist System

As you can see in the robot's block diagram, the TOMBOT is very much a minimalist system with just enough components to demonstrate autonomous behaviors: Cruise, Escape, Avoid, and Home. All these behaviors require the use of left and right servos for autonomous maneuverability.

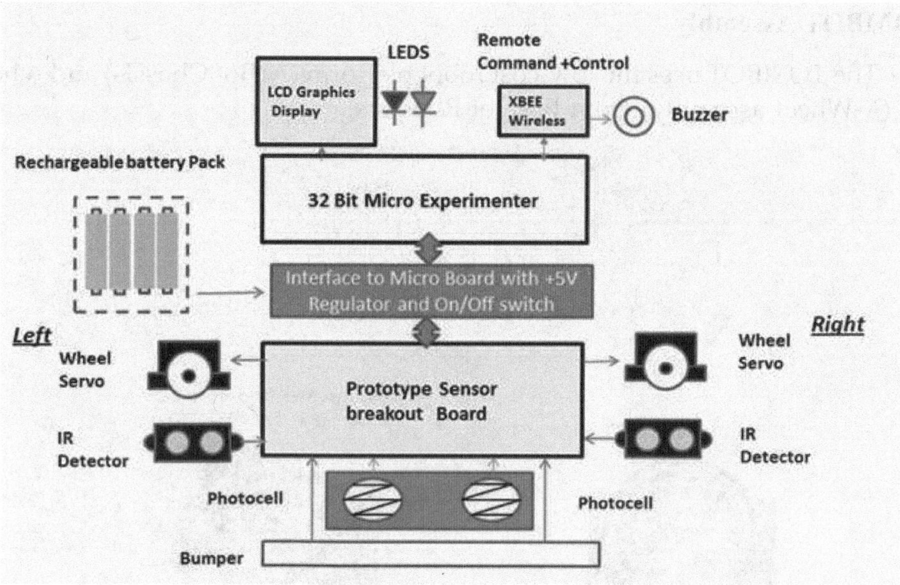

Fig. The TOMBOT system

The Cruise behavior just keeps the robot in motion in lieu of any stimulus. The Escape behavior uses the bumper to sense a collision and then 180° spin with reverse.

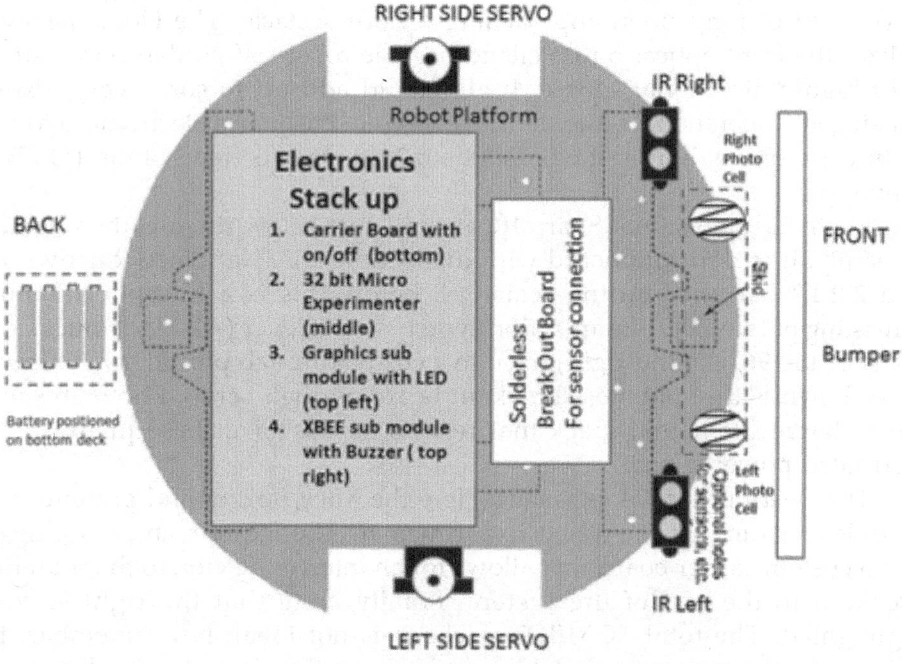

Fig. Basic sensor and electronic layout

TOMBOT Assembly

The TOMBOT uses the low-cost robot platform (ArBot Chassis) and wheel set (X-Wheel assembly) from Budget Robotics.

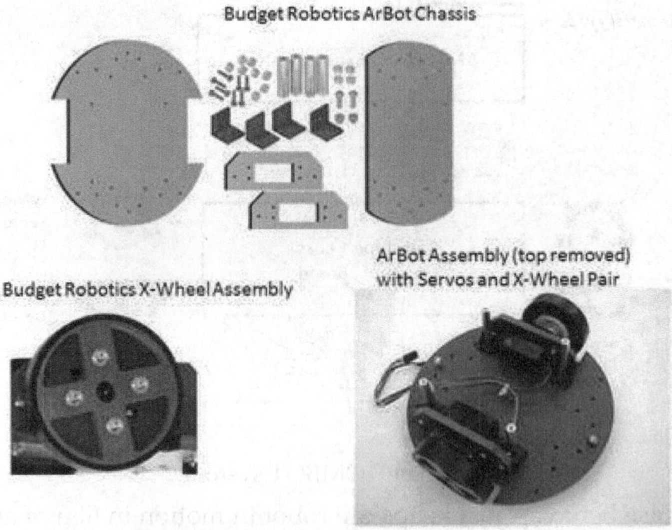

Fig. The platform and wheel set

The Avoid behavior makes use of continuous forward-looking IR sensors to veer left or right upon approaching a close obstacle. The Home behavior utilizes the front optical photocells to provide robot self-guidance to a strong light highly directional source. It all should add up to some very distinct "intelligent" operation. Figure depicts the basic sensor and electronic layout. A picture is worth a thousand words. Photo 2 shows two views of the TOMBOT prototype.

Photo 2a shows dual Sharp IR sensors. Just below them is the photocell assembly. It is a custom board with dual CdS GL5528 photoconductive cells and 2.2-k? current-limiting resistors. Below this is a bumper assembly consisting of two SPDT Snap-action switches with lever (All Electronics Corp. CAT# SMS-196, left and right) fixed to a custom pre-fab plastic front bumper. Also shown is the solderless breakout board and left servo. Photo 2b shows the rechargeable battery pack that resides on the lower base platform and associated power switch.

The electronics stack is visible. Here the XBee/Buzzer and graphics card modules residing on the 32-bit Experimenter. The Experimenter is plugged into a custom carrier board that allows for an interconnection to the solderless breakout to the rest of the system. Finally, note that the right servo is highlighted. The total TOMBOT package is not ideal; but remember, I'm talking about a prototype, and this particular configuration has held up nicely in several field demos.

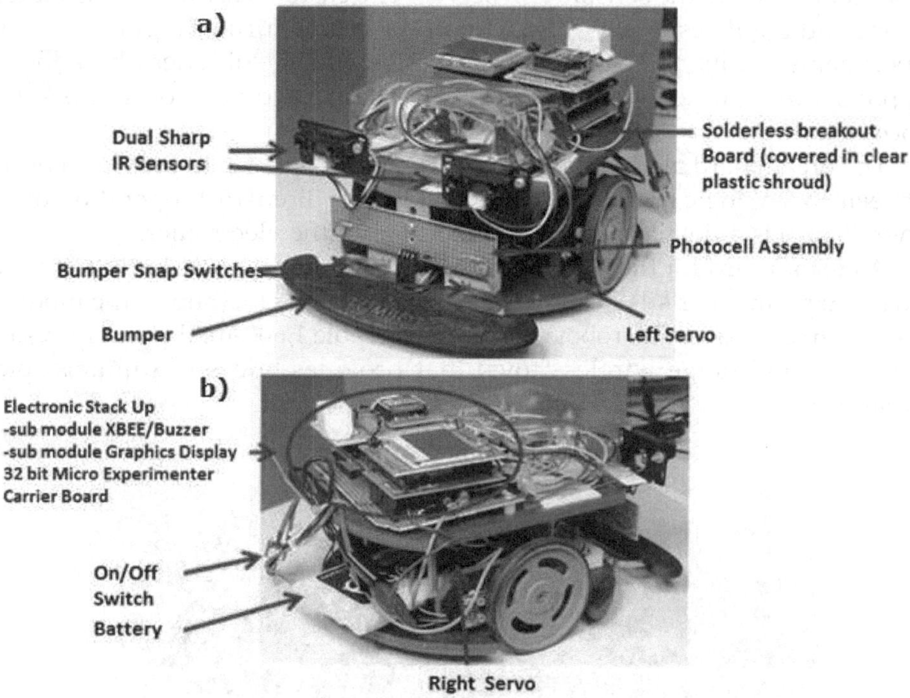

Fig. Photo 2a: The TOMBOT's Sharp IR sensors, photo assembly, and more. b: The battery pack, right servo, and more.

I used Parallax (Futaba) continuous-rotation servos. They use a three-wire connector (+5 V, GND, and Control).

Figure depicts a second-generation bumper assembly. The same snap-action switches with extended levers are bent and fashioned to interconnect a bumper assembly as shown.

TOMBOT Electronics

A 32-bit Micro Experimenter is used as the CPU. This board is based the high-end Microchip Technology PIC32MX695F512H 64-pin TQFP with 128-KB RAM, 512-KB flash memory, and an 80-MHz clock. I did not want to skimp on this component during the prototype phase. In addition the 32-bit Experimenter supports a 102 × 64 monographic card with green/red backlight controls and LEDs. Since a full graphics library was already bundled with this Experimenter graphics card, it also represented good risk reduction during prototyping phase. Details for both cards are available on theKiba website.

The Experimenter supports six basic board-level connections to outside world using JP1, JP2, JP3, JP4, BOT, and TOP headers. A custom carrier board interfaces to the Experimenter via these connections and provides power and signal connection to the sensors and servos. The custom carrier accepts battery voltage and regulates it to +5 VDC. This +5 V is then further regulated by the Experimenter to its native +3.3-VDC operation. The solderless breadboard supports a resistor network to sense a +9-V battery voltage for a +3.3-V PIC processor.

The breadboard also contains an LM324 quad op-amp to provide a buffer between +3.3-V logic of the processor and the required +5-V operation of the servo. Figure is a detailed schematic diagram of the electronics.

A custom card for the XBee radio carrier and buzzer was built that plugs into the Experimenter's TOP and BOT connections. Photo 3 shows the modules and the carrier board. The robot uses a rechargeable 1,600-mAH battery system (typical of mid-range wireless toys) that provides hours of uninterrupted operation.

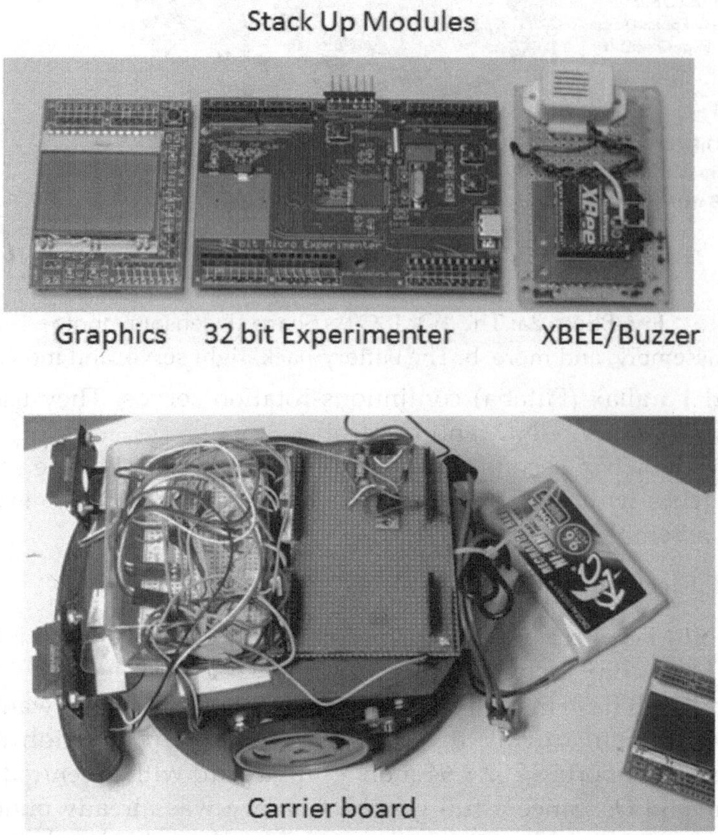

Fig. The modules and the carrier board

PIC32 On-Chip Peripherals

The major PIC32 peripheral connection for the Experimenter to rest of the system is shown. The TOMBOT uses PWM for servo, UART for XBee, SPI and digital for LCD, analog input channels for all the sensors, and digital for the buzzer and bumper detect. The key peripheral connection for the Experimenter to rest of the system is shown in Figure. The PIC32 pinouts and their associated Experimenter connections are detailed in Figure.

PIC32 Peripheral Connections

- **ADC Converter Channels**
  - AN5 Left IR Sensor JP2 Pin 7
  - AN4 Right IR Sensor JP2 Pin 8
  - AN8 Battery Power JP4 Pin 8
  - AN9 Left Photo Cell BOT Pin 1
  - AN10 Right Photo Cell BOT Pin 2
- **PWM outputs**
  - OC3 Right Wheel Servo JP2 Pin 1
  - OC4 Left Wheel Servo JP2 Pin 2
- **Digital I/O**
  - RG9 Left Bumper In JP4 Pin 3
  - RG6 Right Bumper In JP4 Pin 5
  - RF1- Buzzer Out- TOP Pin 7
  - RD5 – XBEE Reset- BOT Pin 7
- **Uart1B**
  - U1BRX -XBEE DOUT-BOT Pin 5
  - U1BTX- XBEE DIN-BOT Pin 7
- **LCD Graphics Display**
  - RF4 Red LED JP3 Pin 5
  - Rd10 Green LED JP4 Pin 1
  - RB0 Chip Select JP3 Pin 1
  - RB3 Red LED Backlight JP3 Pin 4
  - RD4 Green LED Backlight JP4 Pin 2
  - RB1 Reset JP3 Pin 2
  - RB2 Command/Data JP3 Pin 3
  - SCK2A Clock JP4 Pin 6
  - SD02A Data IN JP4 Pin 4

Fig. PIC32 peripheral pinouts and EXP32 connectors

The TOMBOT Motion Basics and the PIC32 Output Compare Peripheral

Let's review the basics for TOMBOT motor control. The servos use the Parallax (Futaba) Continuous Rotation Servos. With two-wheel control, the robot motion is controlled as per Table.

Table. Robot motion

Robot Motion	Left Servo	Right Servo
FORWARD	CCW	CW
BACKWARD	CW	CCW
LEFT TURN	CW	CW
RIGHT TURN	CCW	CCW
PIVOT LEFT	CENTER	CW
PIVOT RIGHT	CCW	CENTER
STOP	CENTER	CENTER

The servos are controlled by using a 20-ms (500-Hz) pulse PWM pattern where the PWM pulse can from 1.0 ms to 2.0 ms. The effects on the servos for the different PWM are shown in Figure.

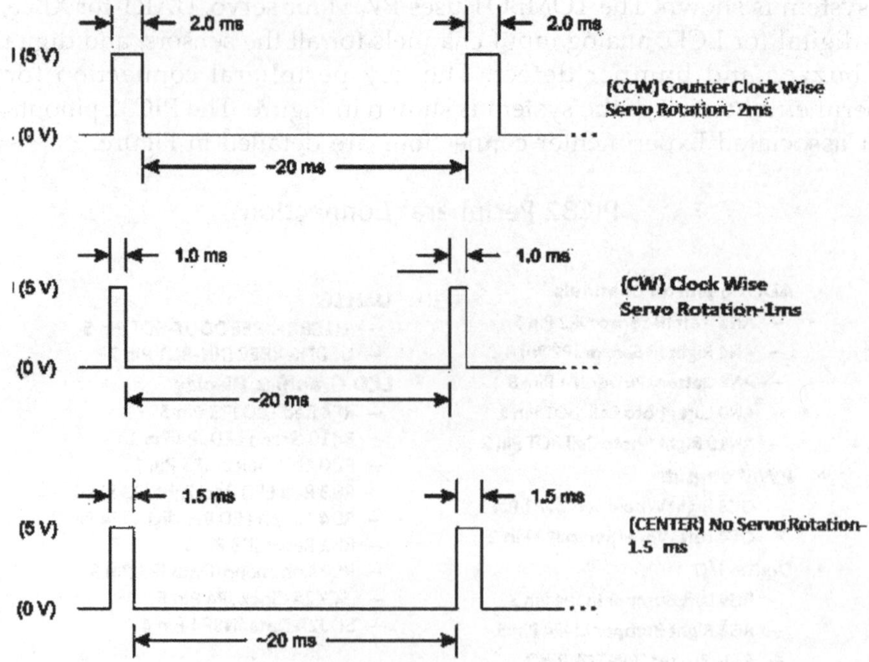

Fig. Servo PWM control

The PIC32 microcontroller (used in the Experimenter) has five Output Compare modules (OCX, where X =1, 2, 3, 4, 5).

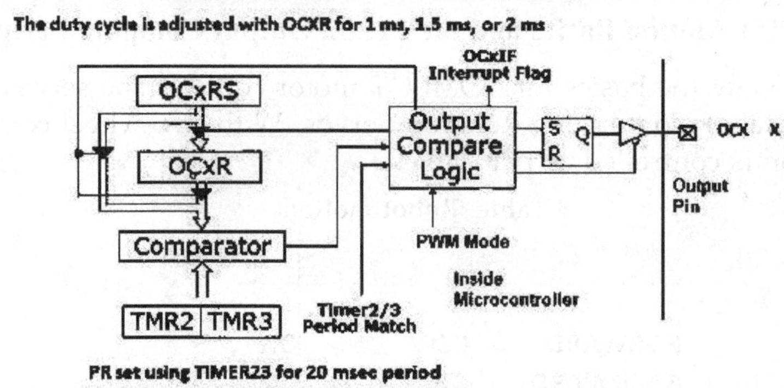

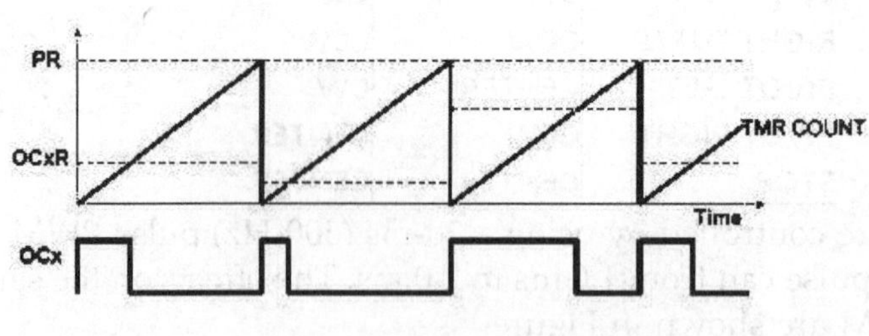

Fig. PWM generation

We use two of these peripherals, specifically OC3, OC4 to generate the PWM to control the servo speed and direction.

The OCX module can use either 16 Timer2 (TMR2) or 16 Timer3 (TMR3) or combined as 32-bit Timer23 as a time base and for period (PR) setting for the output pulse waveform. In our case, we are using Timer23 as a PR set to 20 ms (500 Hz). The OCXRS and OCXR registers are loaded with a 16-bit value to control width of the pulse generated during the output period. This value is compared against the Timer during each period cycle. The OCX output starts high and then when a match occurs OCX logic will generate a low on output. This will be repeated on a cycle-by-cycle basis.

HOW AUTONOMOUS ROBOTS WORK

ROBOT BASICS

The vast majority of robots do have several qualities in common. First of all, almost all robots have a movable body. Some only have motorized wheels, and others have dozens of movable segments, typically made of metal or plastic. Like the bones in your body, the individual segments are connected together with joints.

Robots spin wheels and pivot jointed segments with some sort ofactuator. Some robots use electric motors and solenoids as actuators; some use a hydraulic system; and some use a pneumatic system (a system driven by compressed gases). Robots may use all these actuator types.

A robot needs a power source to drive these actuators. Most robots either have a battery or they plug into the wall. Hydraulic robots also need a pump to pressurize the hydraulic fluid, and pneumatic robots need an air compressor or compressed air tanks.

Fig. NASA's Urbie climbing stairs

The actuators are all wired to an electrical circuit. The circuit powers electrical motors and solenoids directly, and it activates the hydraulic system by manipulating electrical valves. The valves determine the pressurized fluid's path through the machine. To move a hydraulic leg, for example, the robot's controller would open the valve leading from the fluid pump to a piston cylinder attached to that leg. The pressurized fluid would extend the piston, swiveling the leg forward. Typically, in order to move their segments in two directions, robots use pistons that can push both ways.

The robot's computer controls everything attached to the circuit. To move the robot, the computer switches on all the necessary motors and valves. Most robots arereprogrammable -- to change the robot's behavior, you simply write a new program to its computer.

Not all robots have sensory systems, and few have the ability to see, hear, smell or taste. The most common robotic sense is the sense of movement -- the robot's ability to monitor its own motion. A standard design uses slotted wheels attached to the robot's joints. An LED on one side of the wheel shines a beam of light through the slots to a light sensor on the other side of the wheel. When the robot moves a particular joint, the slotted wheel turns. The slots break the light beam as the wheel spins. The light sensor reads the pattern of the flashing light and transmits the data to the computer. The computer can tell exactly how far the joint has swiveled based on this pattern. This is the same basic system used in computer mice.

These are the basic nuts and bolts of robotics. Roboticists can combine these elements in an infinite number of ways to create robots of unlimited complexity. In the next section, we'll look at one of the most popular designs, the robotic arm.

The Robotic Arm

The term robot comes from the Czech word robota, generally translated as "forced labor." This describes the majority of robots fairly well. Most robots in the world are designed for heavy, repetitive manufacturing work. They handle tasks that are difficult, dangerous or boring to human beings.

The most common manufacturing robot is the robotic arm. A typical robotic arm is made up of seven metal segments, joined by six joints. The computer controls the robot by rotating individual step motors connected to each joint (some larger arms use hydraulics or pneumatics). Unlike ordinary motors, step motors move in exact increments (check out Anaheim Automation to find out how). This allows the computer to move the arm very precisely, repeating exactly the same movement over and over again. The robot uses motion sensors to make sure it moves just the right amount.

An industrial robot with six joints closely resembles a human arm -- it has the equivalent of a shoulder, an elbow and a wrist. Typically, the shoulder is mounted to a stationary base structure rather than to a movable body. This type of robot has six degrees of freedom, meaning it can pivot in six different ways. A human arm, by comparison, has seven degrees of freedom.

Your arm's job is to move your hand from place to place. Similarly, the robotic arm's job is to move an end effector from place to place. You can outfit robotic arms with all sorts of end effectors, which are suited to a particular application. One common end effector is a simplified version of the hand, which can grasp and carry different objects. Robotic hands often have built-in pressure sensors that tell the computer how hard the robot is gripping a particular object. This keeps the robot from dropping or breaking whatever it's carrying. Other end effectors include blowtorches, drills and spray painters.

Industrial robots are designed to do exactly the same thing, in a controlled environment, over and over again. For example, a robot might twist the caps onto peanut butter jars coming down an assembly line. To teach a robot how to do its job, the programmer guides the arm through the motions using a handheld controller. The robot stores the exact sequence of movements in its memory, and does it again and again every time a new unit comes down the assembly line.

Most industrial robots work in auto assembly lines, putting cars together. Robots can do a lot of this work more efficiently than human beings because they are so precise. They always drill in the exactly the same place, and they always tighten bolts with the same amount of force, no matter how many hours they've been working. Manufacturing robots are also very important in the computer industry. It takes an incredibly precise hand to put together a tiny microchip.

Mobile Robots

Robotic arms are relatively easy to build and program because they only operate within a confined area. Things get a bit trickier when you send a robot out into the world.

Fig. Fujitsu's HOAP-1 robot

The first obstacle is to give the robot a working locomotion system. If the robot will only need to move over smooth ground, wheels or tracks are the best option. Wheels and tracks can also work on rougher terrain if they are big enough. But robot designers often look to legsinstead, because they are more adaptable. Building legged robots also helps researchers understand natural locomotion -- it's a useful exercise in biological research.

Typically, hydraulic or pneumatic pistons move robot legs back and forth. The pistons attach to different leg segments just likemuscles attach to different bones. It's a real trick getting all these pistons to work together properly. As a baby, your brain had to figure out exactly the right combination of muscle contractions to walk upright without falling over. Similarly, a robot designer has to figure out the right combination of piston movements involved in walking and program this information into the robot's computer. Many mobile robots have a built-in balance system (a collection of gyroscopes, for example) that tells the computer when it needs to correct its movements.

Fig. NASA's Frogbot uses springs, linkages and motors to hop from place to place.

Bipedal locomotion (walking on two legs) is inherently unstable, which makes it very difficult to implement in robots. To create more stable robot walkers, designers commonly look to the animal world, specifically insects. Six-legged insects have exceptionally good balance, and they adapt well to a wide variety of terrain.

Some mobile robots are controlled by remote -- a human tells them what to do and when to do it. The remote control might communicate with the robot through an attached wire, or using radio or infrared signals. Remote robots, often called puppet robots, are useful for exploring dangerous or inaccessible environments, such as the deep sea or inside a volcano. Some robots are only partially controlled by remote. For example, the operator might direct the robot to go to a certain spot, but not steer it there -- the robot would find its own way.

Autonomous Robots

Autonomous robots can act on their own, independent of any controller. The basic idea is to program the robot to respond a certain way to outside stimuli. The very simple bump-and-go robotis a good illustration of how this works.

This sort of robot has a bumper sensor to detect obstacles. When you turn the robot on, it zips along in a straight line. When it finally hits an obstacle, the impact pushes in its bumper sensor. The robot's programming tells it to back up, turn to the right and move forward again, in response to every bump. In this way, the robot changes direction any time it encounters an obstacle.

Advanced robots use more elaborate versions of this same idea. Roboticists create new programs and sensor systems to make robots smarter and more perceptive. Today, robots can effectively navigate a variety of environments.

Simpler mobile robots use infrared or ultrasound sensors to see obstacles. These sensors work the same way as animal echolocation: The robot sends out a sound signal or a beam of infrared light and detects the signal's reflection. The robot locates the distance to obstacles based on how long it takes the signal to bounce back.

Urbie's view

More advanced robots use stereo vision to see the world around them. Two cameras give these robots depth perception, and image-recognition software gives them the ability to locate and classify various objects. Robots might also use microphones and smellsensors to analyze the world around them.

Some autonomous robots can only work in a familiar, constrained environment. Lawn-mowing robots, for example, depend on buried border markers to define the limits of their yard. An office-cleaning robot might need a map of the building in order to maneuver from point to point. More advanced robots can analyze and adapt to unfamiliar environments, even to areas with rough terrain. These robots may associate certain terrain patterns with certain actions. A rover robot, for example, might construct a map of the land in front of it based on its visual sensors. If the map shows a very bumpy terrain pattern, the robot knows to travel another way. This sort of system is very useful for exploratory robots that operate on other planets (check out JPL Robotics to learn more).

An alternative robot design takes a less structured approach -- randomness. When this type of robot gets stuck, it moves its appendages every which way until something works. Force sensors work very closely with the actuators, instead of the computer directing everything based on a program. This is something like an ant trying to get over an obstacle -- it doesn't seem to make a decision when it needs to get over an obstacle, it just keeps trying things until it gets over it.

Homebrew Robots

In the last couple of sections, we looked at the most prominent fields in the world of robots -- industry robotics and research robotics. Professionals in these fields have made most of the major advancements in robotics over the years, but they aren't the only ones making robots. For decades, a small but passionate band of hobbyists has been creating robots in garages and basements all over the world.

Homebrew robotics is a rapidly expanding subculture with a sizable Web presence. Amateur roboticists cobble together their creations using commercial robot kits, mail order components, toys and even oldVCRs.

Homebrew robots are as varied as professional robots. Some weekend roboticists tinker with elaborate walking machines, some design their own service bots and others create competitive robots. The most familiar competitive robots are remote control fighters like you might see on "BattleBots." These machines aren't considered "true robots" because they don't have reprogrammable computer brains. They're basically souped-up remote control cars.

More advanced competitive robots are controlled by computer. Soccer robots, for example, play miniaturized soccer with no human input at all. A standard soccer bot team includes several individual robots that communicate with a central computer. The computer "sees" the entire soccer field with a video camera and picks out its own team members, the opponent's members, the ball and the goal based on their color. The computer processes this information at every second and decides how to direct its own team.

Check out the official RoboCup Web site for more information on Soccer robots, and Google > Computers > Robotics > Competitions for information on other robot competitions. Google > Computers > Robotics > Building will give you more information on building your own robots.

Adaptable and Universal

The personal computer revolution has been marked by extraordinary adaptability. Standardized hardware and programming languages let computer engineers and amateur programmers mold computers to their own particular purposes. Computer components are sort of like art supplies -- they have an infinite number of uses.

Most robots to date have been more like kitchen appliances. Roboticists build them from the ground up for a fairly specific purpose. They don't adapt well to radically new applications.

This situation may be changing. A company called Evolution Robotics is pioneering the world of adaptable robotics hardware and software. The company hopes to carve out a niche for itself with easy-to-use "robot developer kits."

The kits come with an open software platform tailored to a range of common robotic functions. For example, roboticists can easily give their creations the ability to follow a target, listen to voice commands and maneuver around obstacles. None of these capabilities are revolutionary from a technology standpoint, but it's unusual that you would find them in one simple package.

The kits also come with common robotics hardware that connects easily with the software. The standard kit comes with infrared sensors, motors, a microphone and a video camera. Roboticists put all these pieces together with a souped-up erector set -- a collection of aluminum body pieces and sturdy wheels.

These kits aren't your run-of-the-mill construction sets, of course. At upwards of $700, they're not cheap toys. But they are a big step toward a new sort of robotics. In the near future, creating a new robot to clean your house or take care of your pets while you're away might be as simple as writing a BASIC program to balance your checkbook.

ROBOTS AND ARTIFICIAL INTELLIGENCE

Artificial intelligence (AI) is arguably the most exciting field in robotics. It's certainly the most controversial: Everybody agrees that a robot can work in an assembly line, but there's no consensus on whether a robot can ever be intelligent.

Like the term "robot" itself, artificial intelligence is hard to define. Ultimate AI would be a recreation of the human thought process -- a man-made machine with our intellectual abilities. This would include the ability to learn just about anything, the ability to reason, the ability to use language and the ability to formulate original ideas. Roboticists are nowhere near achieving this level of artificial intelligence, but they have made a lot of progress with more limited AI. Today's AI machines can replicate some specific elements of intellectual ability. Computers can already solve problems in limited realms. The basic idea of AI problem-solving is very simple, though its execution is complicated. First, the AI robot or computer gathers facts about a situation through sensors or human input.

The computer compares this information to stored data and decides what the information signifies. The computer runs through various possible actions and predicts which action will be most successful based on the collected information. Of course, the computer can only solve problems it's programmed to solve -- it doesn't have any generalized analytical ability. Chess computers are one example of this sort of machine.

Some modern robots also have the ability to learn in a limited capacity. Learning robots recognize if a certain action (moving its legs in a certain way, for instance) achieved a desired result (navigating an obstacle). The robot stores this information and attempts the successful action the next time it encounters the same situation. Again, modern computers can only do this in very limited situations. They can't absorb any sort of information like a human can. Some robots can learn by mimicking human actions. In Japan, roboticists have taught a robot to dance by demonstrating the moves themselves.

Some robots can interact socially. Kismet, a robot at M.I.T's Artificial Intelligence Lab, recognizes human body language and voice inflection and responds appropriately. Kismet's creators are interested in how humans and babies interact, based only on tone of speech and visual cue. This low-level interaction could be the foundation of a human-like learning system.

Kismet and other humanoid robots at the M.I.T. AI Lab operate using an unconventional control structure. Instead of directing every action using a central computer, the robots control lower-level actions with lower-level computers. The program's director, Rodney Brooks, believes this is a more accurate model of human intelligence. We do most things automatically; we don't decide to do them at the highest level of consciousness.

The real challenge of AI is to understand how natural intelligence works. Developing AI isn't like building an artificial heart -- scientists don't have a simple, concrete model to work from. We do know that the braincontains billions and billions of neurons, and that we think and learn by establishing electrical connections between different neurons. But we don't know exactly how all of these connections add up to higher reasoning, or even low-level operations. The complex circuitry seems incomprehensible.

Because of this, AI research is largely theoretical. Scientists hypothesize on how and why we learn and think, and they experiment with their ideas using robots. Brooks and his team focus on humanoid robots because they feel that being able to experience the world like a human is essential to developing human-like intelligence.

It also makes it easier for people to interact with the robots, which potentially makes it easier for the robot to learn.

Just as physical robotic design is a handy tool for understanding animal and human anatomy, AI research is useful for understanding how natural intelligence works. For some roboticists, this insight is the ultimate goal of designing robots. Others envision a world where we live side by side with intelligent machines and use a variety of lesser robots for manual labor, health care and communication. A number of robotics experts predict that robotic evolution will ultimately turn us into cyborgs -- humans integrated with machines. Conceivably, people in the future could load their minds into a sturdy robot and live for thousands of years!

In any case, robots will certainly play a larger role in our daily lives in the future. In the coming decades, robots will gradually move out of the industrial and scientific worlds and into daily life, in the same way that computers spread to the home in the 1980s.

The best way to understand robots is to look at specific designs. The links on the next page will show you a variety of robot projects around the world.

3

Mobile Robot Localization and Mapping

SIMULTANEOUS LOCALIZATION AND MAPPING

Fig. A robot built by Technische Universität Darmstadt maps a maze using a LIDAR.

Fig. A map generated by the Darmstadt Robot.

Simultaneous localization and mapping (SLAM) is a technique used by robots and autonomous vehicles to build up a map within an unknown environment (without a priori knowledge), or to update a map within a known environment (with a prioriknowledge from a given map), while at the same time keeping track of their current location.

OPERATIONAL DEFINITION

Maps are used to determine a location within an environment and to depict an environment for planning and navigation; they support the assessment of actual location by recording information obtained from a form of perception and comparing it to a current set of perceptions. The benefit of a map in aiding the assessment of a location increases as the precision and quality of the current perceptions decrease. Maps generally represent the state at the time that the map is drawn; this is not necessarily consistent with the state of the environment at the time the map is used.

The complexity of the technical processes of locating and mapping under conditions of errors and noise do not allow for a coherent solution of both tasks. Simultaneous localization and mapping (SLAM) is a concept that binds these processes in a loop and therefore supports the continuity of both aspects in separated processes; iterative feedback from one process to the other enhances the results of both consecutive steps.

Mapping is the problem of integrating the information gathered by a set of sensors into a consistent model and depicting that information as a given representation. It can be described by the first characteristic question, What does the world look like? Central aspects in mapping are the representation of the environment and the interpretation of sensor data.

In contrast to this, localization is the problem of estimating the place (and pose) of the robot relative to a map; in other words, the robot has to answer the second characteristic question, Where am I? Typically, solutions comprise tracking, where the initial place of the robot is known, and global localization, in which no or just some a priori knowledge of the environmental characteristics of the starting position is given.

SLAM is therefore defined as the problem of building a model leading to a new map, or repetitively improving an existing map, while at the same time localizing the robot within that map. In practice, the answers to the two characteristic questions cannot be delivered independently of each other.

Before a robot can contribute to answering the question of what the environment looks like, given a set of observations, it needs to know e.g.:

- The robot's own kinematics,
- Which qualities the autonomous acquisition of information has, and,
- From which sources additional supporting observations have been made.

It is a complex task to estimate the robot's current location without a map or without a directional reference. "Location" may refer to simply the position of the robot or might also include its orientation.

TECHNICAL PROBLEMS

SLAM can be thought of as a chicken or egg problem: An unbiased map is needed for localization while an accurate pose estimate is needed to build that map. This is the starting condition for iterative mathematical solution strategies.

Beyond, the answering of the two characteristic questions is not as straightforward as it might sound due to inherent uncertainties in discerning the robot's relative movement from its various sensors. Generally, due to the budget of noise in a technical environment, SLAM is not served with just compact solutions, but with a bunch of physical concepts contributing to results.

If at the next iteration of map building the measured distance and direction traveled has a budget of inaccuracies, driven by limited inherent precision of sensors and additional ambient noise, then any features being added to the map will contain corresponding errors. Over time and motion, locating and mapping errors build cumulatively, grossly distorting the map and therefore the robot's ability to determine its actual location and heading with sufficient accuracy.

There are various techniques to compensate for errors, such as recognizing features that it has come across previously (i.e., data association or loop closure detection), and re-skewing recent parts of the map to make sure the two instances of that feature become one. Statistical techniques used in SLAM include Kalman filters, particle filters (aka.Monte Carlo methods) and scan matching of range data. They provide an estimation of the posterior probability function for the pose of the robot and for the parameters of the map. Set-membership techniques are mainly based on interval constraint propagation. They provide a set which encloses the pose of the robot and a set approximation of the map.

Mapping

SLAM in the mobile robotics community generally refers to the process of creating geometrically consistent maps of the environment. Topological maps are a method of environment representation which capture the connectivity (i.e., topology) of the environment rather than creating a geometrically accurate map. As a result, algorithms that create topological maps are not referred to as SLAM.

SLAM is tailored to the available resources, hence not aimed at perfection, but at operational compliance. The published approaches are employed in unmanned aerial vehicles, autonomous underwater vehicles, planetary rovers, newly emerging domestic robots and even inside the human body.

It is generally considered that "solving" the SLAM problem has been one of the notable achievements of the robotics research in the past decades. The related problems ofdata association and computational complexity are among the problems yet to be fully resolved. A significant recent advance in the feature based SLAM literature involved the re-examination of the probabilistic foundation for Simultaneous Localisation and Mapping (SLAM) where it was posed in terms of multi-object Bayesian filtering with random finite sets that provide superior performance to leading feature-based SLAM algorithms in challenging measurement scenarios with high false alarm rates and high missed detection rates without the need for data association.

Sensing

SLAM will always use several different types of sensors to acquire data with statistically independent errors. Statistical independence is the mandatory requirement to cope with metric bias and with noise in measures.

Such optical sensors may be one dimensional (single beam) or 2D-(sweeping) laser rangefinders, 3D Flash LIDAR, 2D or 3D sonar sensors and one or more 2D cameras.

Since 2005, there has been intense research into VSLAM (visual SLAM) using primarily visual (camera) sensors, because of the increasing ubiquity of cameras such as those in mobile devices.

Recent approaches apply quasi-optical wireless ranging for multi-lateration (RTLS) or multi-angulation in conjunction with SLAM as a tribute to erratic wireless measures.

A special kind of SLAM for human pedestrians uses a shoe mounted inertial measurement unit as the main sensor and relies on the fact that pedestrians are able to avoid walls. This approach called FootSLAM can be used to automatically build floor plans of buildings that can then be used by an indoor positioning system.

Locating

The results from sensing will feed the algorithms for locating. According to propositions of geometry, any sensing must include at least one lateration and (n+1) determining equations for an n-dimensional problem. In addition, there must be some additional a priori knowledge about orienting the results versus absolute or relative systems of coordinates with rotation and mirroring.

Modeling

Contribution to mapping may work in 2D modeling and respective representation or in 3D modeling and 2D projective representation as well. As a part of the model, the kinematics of the robot is included, to improve estimates of sensing under conditions of inherent and ambient noise. The dynamic model balances the contributions from various sensors, various partial error models and finally comprises in a sharp virtual depiction as a map with the location and heading of the robot as some cloud of probability. Mapping is the final depicting of such model, the map is either such depiction or the abstract term for the model.

LITERATURE

A seminal work in SLAM is the research of R.C. Smith and P. Cheeseman on the representation and estimation of spatial uncertainty in 1986. Other pioneering work in this field was conducted by the research group of Hugh F. Durrant-Whyte in the early 1990s.

BIOLOGICALLY INSPIRED MACHINES: MAPPING AND LOCALIZATION

The task of building a robot capable of concurrently mapping and localizing (CML) in an unexplored and dynamic environment is an outstanding problem in the field of robotic autonomy. Although there are implementations of CML that have solved the problem, most have relied on small environments or time and memory constraints that are not be representative of the real world. For example, robots using an Extended Kalman Filter (EKF) are able to correctly perform CML, but are unable to handle the Relocation Problem, when the robot is kidnapped and moved and must then continue running CML on the original map. Different implementations that seek to handle Relocation often cannot work in real-time, or rely on an extensive memory, limiting the size of the map that they can handle [1].

While most research to the problems in CML lies in further enhancements of the existing algorithms, this paper looks for inspiration in the natural world. Human mechanisms for handling the relocation problem [2] and limited memory capacity [3] are real-time functions. These models of human memory can be drawn analogously into an existing implementation of CML, and provide the robustness that is currently lacking.

This paper describes an existing implementation of Simultaneous Localization and Mapping (SLAM) that is capable, but fundamentally limited. The existing implementation can only handle a small number of features in the environment and is unable to handle the relocation problem. The two biologically inspired enhancements to the SLAM implementation provide the following new capabilities:

- Handle the relocation problem. If the robot is abruptly moved to a new place in the map, it undergoes a real-time, sample-based search of its map to determine its new position and orientation before continuing its plan. This mirrors the biological recognition system that allows humans to recognize and readjust to places they have previously seen.
- Handle larger environments with memory management. As the robot explores, certain stored features in the map will slowly decay until they reach a threshold, when they are forgotten. As a result, the robot keeps sparser maps of areas it has not seen in a while,

and more detailed maps of more recent areas. This allows the robot to maintain old information, while still adding to the map. This mirrors the biological memory storage system where humans remember old facts, but not with the same depth and richness as more current information.

Using these two biological systems, a simulated running of the enhanced SLAM implementation is able to handle being kidnapped as well as environments with many times more features.

The original SLAM implementation, both biological improvements and experimentation that shows the added robustness of the new features. The SLAM implementation, including how observations were made and stored, as well as how it solved the basic CML problem. The descriptions of the Recognition and Memory Storage enhancements described briefly above. Our experimental results after running the robot on a simulated environment with and without the biological enhancements.

SLAM IMPLEMENTATION

The initial SLAM implementation used a system of linearized stochastic mapping to handle the problem of CML [6]. The robot was given a movement plan for exploration and at each step along the way, took observations of the environment with a limited range scan. Observations were matched to features and the robot updated its map and localized itself using a Kalman Filter.

The actual task of observation and preprocessing the feature information is dependent on an artificially simplified simulation environment. Each feature is given a unique identifier. When the robot scans for observations, it measures the distances to a set of features within its scan range, with some arbitrary error adjustment to simulate real-world noise. However, it also receives the unique identifier of each feature, which is unrealistic in a real world setting, but adequate for the simulated intents of this SLAM implementation. The robot compare the feature ID against its stored map, and determines whether or not it is a new feature. If it is in fact a new feature, it is added to the map. If it is not, the robot has to update the map.

The heart of stochastic mapping lies in the second stage where the features are actually processed. This implementation relies on a Kalman Filter Update to handle the necessary computation of updating the map and localizing the robot. The KF approach relies on creating a Gaussian distribution of the predicted state of the environment. Using the measurements from a feature observation, a Bayes filter is used to correct the prediction and update the state [6].

Using a simple loop of motion, prediction, observation, update, the SLAM algorithm is able to successfully map out a feature-laden environment. Given the random observation errors, the first-pass map is within 2 standard deviations of the correct map.

RECOGNITION

A memory recognition system that will allow the existing SLAM implementation to handle the relocation problem, when the robot is moved in an unknown region of the map. Since the existing SLAM implementation relied on feature identification, which makes the relocation problem trivial, we ignore feature ID's when observing and processing features in memory recognition. Instead we give additional quantitative value to the feature, an angle of orientation and length that is observable with a degree of error comparable to those in the distance errors. This additional feature information is necessary for narrowing down the choice of possible observable features during recognition.

The recognition technique relies on searching the geometric constraints of the observable features to determine the most likely position and orientation of the robot. In a standard stochastic mapping this search involves matching each observation to a possible feature, and searching through all combinations of observation-feature pairs for the optimal fit. This technique is time inefficient as searching through the feature space is exponential in the number of measurements. This paper draws on a new technique to minimize the search time by limiting the size of the search tree [6].

GEOMETRIC CONSTRAINTS

In order to reduce the size of the observation-feature search tree, we use three different types of geometric constraints to determine the likelihood of an observation-feature pairing. The first of these constraints is a unary test of the feature's angle of orientation with respect to the observed measurement of that angle.

We determine the stochastic parameter vectors and a correlation matrix for the feature and the observed angle. Then, using a chi square test for statistical significance we can decide if the chosen pair should be considered or ruled out.

The second geometric constraint we consider is a binary test comparing the relative geometry's between two feature-observation pairs. In this case we create a vector of the difference between the distances of the features and the observed measurements. Using the corresponding correlation matrix as before, we arrive at a bound of probability for the chi square test.

The final constraint is the joint compatibility test. This test checks the geometric relationship among all the feature-observation pairs. The intent behind this test is to rule out the cases where all subsets of two pairs pass the binary test, but the set, as a whole is geometrically infeasible. The joint computability test computes the chi square test on all the combined pairs. This test makes sure that the inclusion of spurious observations are minimized, since hypotheses with spurious observations will not pass the statistical test.

BRANCH & BOUND

Using the three geometric constraints, we implemented a recursive branch and bound search algorithm that returns the best set of feature-observation pairs. The joint computability branch and bound (JCBB) algorithm works as follows:

- If all observations have been mapped:
- Count the number of feature-observation pairs in the current set, H, and in the previous best set, B.
- Designate the set with more pairs as the new B.
- If the current observation, o, has not yet been mapped to a feature:
- For every feature, fi, in the map:
- If unary(o, fi) AND joint_comp([H, fi]:
- JCBB([H fi])
- If no feature could be mapped to o AND there are more pairs in H than in B:
- JCBB({H 0])//(spurious observation)

Running this algorithm over the set of observations made during attempted recognition is considerably more efficient than the previous attempts to search the entire tree of possible sets of observation-feature pairs. The JCBB algorithm relies on eliminating entire branches as soon as a pair is added which violates the global optimality of the pair set.

RANDOM SAMPLING

Although the JCBB algorithm is a dramatic improvement the full tree search, it still requires a lot of computationally expensive geometric constraint functions. In the case of a move to an unexplored section of the map, the entire search tree still must be explored [6]. One simple heuristic to improve both the worst case and general case search time is to split the search into two parts – a guess and a proof, using a random subset of observations for each part.

When the robot is relocated, it makes n observations of its surroundings. Using the random sample improvement to the JCBB, we select p of these measurements to generate a hypothesis of the new location using only their unary and binary constraints. Without using the expensive joint compatibility test, we reduce the search tree to a significantly smaller hypothesis. Then, we run JCBB on the remaining observations to verify the accuracy of the hypothesis.

MEMORY LIMITATIONS

The biologically inspired memory functions, which allow the SLAM implementation to act more robustly in large environments. The SLAM code as originally implemented was only intended to handle a relatively small number of features. This limit is reflective of two problems in CML in actual robots: (1) limited physical memory, and (2) matching observations to features becomes exponentially harder and more computationally costly as more features are stored. In order to solve these problems, we implemented a memory management system to handle the resources more efficiently.

When a feature is added to our memory, it is assigned a freshness function with probability p. The freshness is equivalent to a time stamp of feature acquisition.

Over time, the freshness will decay and eventually drop below a threshold, at which point it is available to be removed from memory. Similarly, if the feature is observed again, its freshness will return to the initial Peak State. Features will be removed only when an upper limit of features to be held is reached. Furthermore, a feature that is revisited many times will be updated such that it will reflect its robustness in the explored environment.

The result of this new dynamic memory is manifold. Since only a random fraction of features are given freshness functions, the map loses some of its detail, but will retain its general shape. Furthermore, the areas that the robot has most recently explored will be the "freshest" and most densely mapped. Regions that have been mapped in the past, and may not be seen again for a long time, are still kept in memory, much more sparsely.

EXPERIMENTATION

To test the relocation and dynamic memory management system we have used an artificial dataset of an environment similar to a the inside of a building and a vehicle that has an odometry sensor mounted on it. In the beginning of a simulation the vehicle is given a constant velocity and directions as a function of time. The vehicle performs CML while traveling in its environment. Figure depicts concurrent mapping and localization in the simulated environment. The vehicle is able to reliably map the features in its environment and travel through it, compensating for noise contributed by measurements and vehicle controls.

First we tested the relocation implementation. We ran CML continuously until the environment was mapped and then kidnapped the robot. The algorithm was able to map the environment based on very few sensor readings at each step (as few as 3 observations). The ability to map the environment is based mainly on the relationship between the measurements, vehicle position and noise in the system. However, for the recognition to work, the features should demonstrate both intrinsic uniqueness (e.g., whether a feature is a corner or not) and correlations between features that is independent of the vehicle position. That is, the vehicle should be able to uniquely determine its location from a set of features available in its immediate location. Table demonstrates the dramatic effect of distinctiveness on recognition performance.

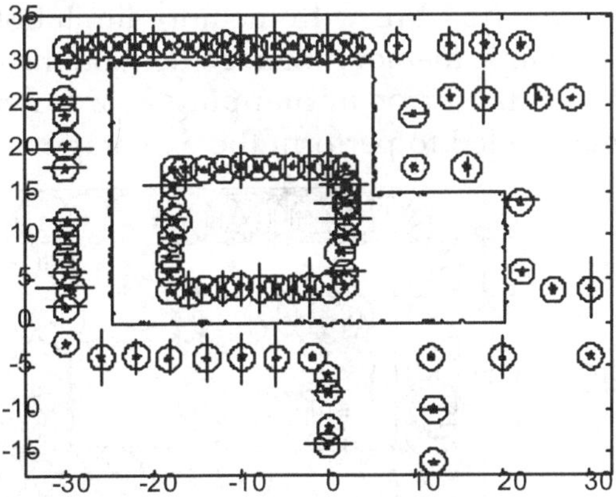

Fig. CML map obtained by the vehicle. The continuous and dashed lines depict the vehicle actual and planned trajectories, respectively. A star marks each feature location, and the associated bar is a unique characteristic of the feature (determined by angle and length). Finally, the error in estimation in vehicle trajectory is marked by the deviation of the continuous line from the dashed line, and the error in estimate the feature location is marked by the radius of the circle.

Changing the length and angle of the bar to correspond to walls and corners simulated distinctiveness of features. For example, all the bars with length close to zero while the other features correspond to walls. Distinctiveness was manipulated by add noise to these characteristics (not shown here). Importantly, the features distinctiveness plays a major role in the successful estimation of the new location.

Table. Relocation estimation error (±SD) as a function of feature distinctiveness. Note that the distinctive features show a much lower error and variance.

	Relocation Estimation Error	
Features	X	y
Normal	-1.30 ± 20.73	-0.58 ± 26.74
Distinct	-0.12 ± .41	0.09 ± 0.39

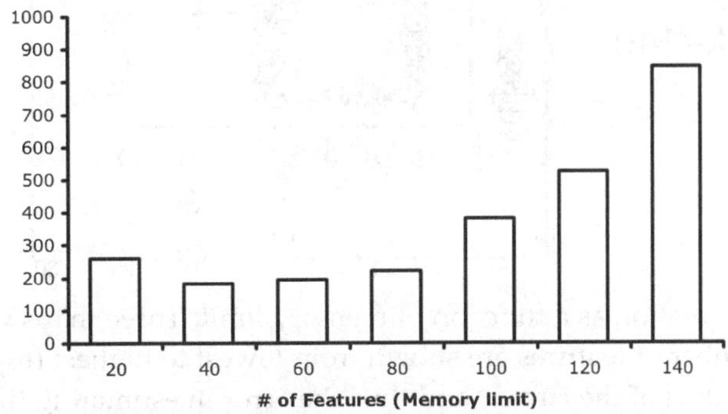

Fig. Mean execution time vs. memory size.

Secondly, we tested the CML with memory limitations. In order to test the effects of using dynamic memory management on the performance of the mapping we measured the error in mapping as a function of number of features and the time needed to perform the exploration.

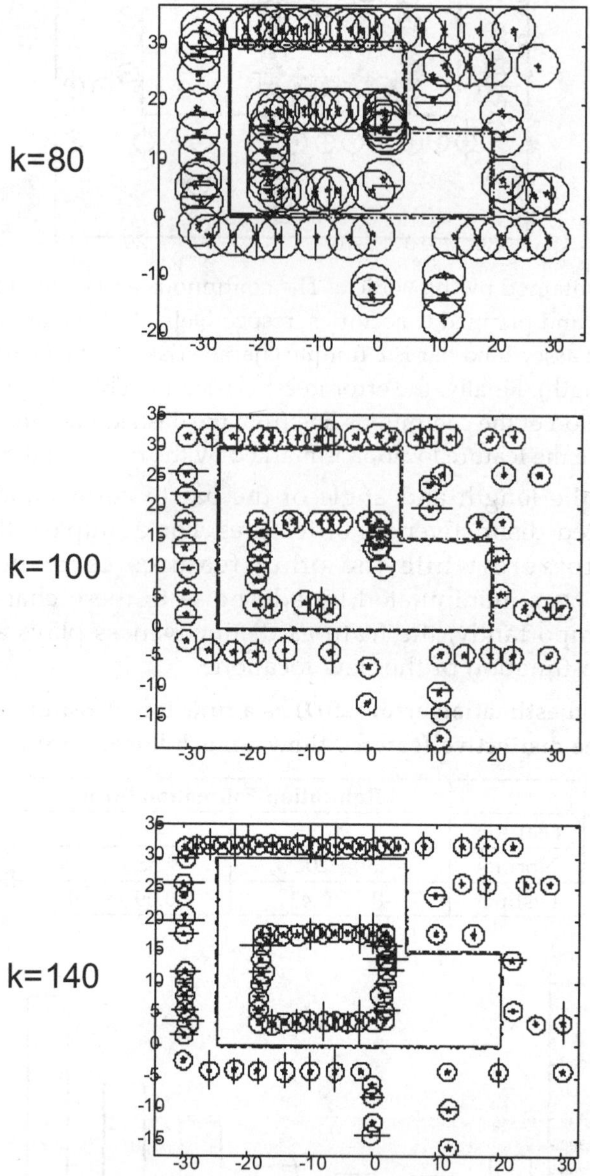

Fig. Feature estimation as a function of memory limit. Three maps with different maximum number of features are shown from lowest to highest (top to bottom). Note that the radius of the circles, and thus the error in estimating their location is inversely related to the memory size.

Figures demonstrate that although the algorithm is successful in mapping the environment with a limit on the number of features and it does so faster then it would do without the memory limitation, it does have larger variance and errors in the resulting map when the memory is limited. Nevertheless, the features' variance does converge in this case and therefore the features do not contribute noise into the mapping and the CML is successful.

Finally, we tested whether the relocation is still effective when using dynamic memory management. The relocation is successful even when memory limitations are applied.

ROBOT LOCOMOTION

Robot locomotion is the collective name for the various methods that robots use to transport themselves from place to place. Although wheeled robots are typically quite energy efficient and simple to control, other forms of locomotion may be more appropriate for a number of reasons (e.g. traversing rough terrain, moving and interacting in human environments). Furthermore, studying bipedal and insect-like robots may beneficially impact on biomechanics.

A major goal in this field is in developing capabilities for robots to autonomously decide how, when, and where to move. However, coordinating a large number of robot joints for even simple matters, like negotiating stairs, is difficult. Autonomous robot locomotion is a major technological obstacle for many areas of robotics, such as humanoids (like Honda's Asimo).

TYPES OF LOCOMOTION

Wheeled

In terms of energy efficiency on flat surfaces, wheeled robots are the most efficient. This is due to the fact that an ideal rolling (but not slipping) wheel loses no energy. A wheel rolling at a given velocity needs no input to maintain its motion. This is in contrast to legged robots which suffer an impact with the ground at heelstrike and lose energy as a result.

There are many different types of wheeled robots, the most common being the Reed Shepps type and the unicycle type. The major concern in the motion planning of wheeled robots are the holonomic that the robot is subject to. These are decided by the type of wheels, number of wheels and the direction of the axes of rotation of the wheels.

Examples:

- iRobot's Roomba
- Various DARPA Grand Challenge entries

Bipedal walking:

- Passive dynamics
- Zero Moment Point

Running:

- ASIMO
- BigDog
- HUBO 2
- RunBot
- Toyota Partner Robot

Rolling

Fig. Segway in the Robot museum in Nagoya.

For simplicity most mobile robots have four wheels or a number of continuous tracks. Some researchers have tried to create more complex wheeled robots with only one or two wheels. These can have certain advantages such as greater efficiency and reduced parts, as well as allowing a robot to navigate in confined places that a four wheeled robot would not be able to.

Examples:

- Boe-Bot
- Cosmobot
- Elmer
- Elsie
- Enon
- HERO
- IRobot Create
- Johns Hopkins Beast
- Land Walker
- Modulus robot
- Musa

- Omnibot
- PaPeRo
- Phobot
- Pocketdelta robot
- Push the Talking Trash Can
- RB5X
- Rovio
- Seropi
- Shakey the robot
- Sony Rolly
- Spykee
- TiLR
- Topo
- TR Araña
- Wakamaru

Hopping

Several robots, built in the 1980s by Marc Raibert at the MIT Leg Laboratory, successfully demonstrated very dynamic walking. Initially, a robot with only one leg, and a very small foot, could stay upright simply by hopping. The movement is the same as that of a person on a pogo stick. As the robot falls to one side, it would jump slightly in that direction, in order to catch itself. Soon, the algorithm was generalised to two and four legs. A bipedal robot was demonstrated running and even performing somersaults.A quadruped was also demonstrated which could trot, run, pace, and bound.

Metachronal motion

Coordinated, sequential mechanical action having the appearance of a traveling wave is called a metachronal rhythm or wave, and is employed in nature by ciliates for transport, and by worms and arthropods for locomotion.

Slithering

Several snake robots have been successfully developed. Mimicking the way real snakes move, these robots can navigate very confined spaces, meaning they may one day be used to search for people trapped in collapsed buildings. The Japanese ACM-R5 snake robot can even navigate both on land and in water.

Examples

- Snake-arm robot
- Roboboa
- Snakebot

Swimming

- Autonomous underwater vehicles

Brachiating

Brachiation allows robots to travel by swinging, only using energy to grab and release surfaces.

ROBOTIC MAPPING

Fig. Robotic mapping can be used for serving robot guide

Robotic mapping is a discipline related to cartography. The goal for an autonomous robot to be able to construct (or use) a map or floor plan and to localize itself in it. Robotic mapping is that branch of one, which deals with the study and application of ability to construct map or floor plan by the autonomous robot and to localize itself in it.

Evolutionarily shaped blind action may suffice to keep some animals alive. For some insects for example, the environment is not interpreted as a map, and they survive only with a triggered response. A slightly more elaborated navigation strategy dramatically enhances the capabilities of the robot. Cognitive maps enable planning capacities, and use of current perceptions, memorized events, and expected consequences.

OPERATION

The robot has two sources of information: the idiothetic and the allothetic sources. When in motion, a robot can use dead reckoning methods such as tracking the number of revolutions of its wheels; this corresponds to the idiothetic source and can give the absolute position of the robot, but it is subject to cumulative error which can grow quickly.

The allothetic source corresponds to the sensors of the robot, like a camera, a microphone, laser, lidar or sonar. The problem here is "perceptual aliasing". This means that two different places can be perceived as the same. For example, in a building, it is nearly impossible to determine a location solely with the visual information, because all the corridors may look the same.

MAP REPRESENTATION

The internal representation of the map can be "metric" or "topological":

- The metric framework is the most common for humans and considers a two-dimensional space in which it places the objects. The objects are placed with precise coordinates. This representation is very useful, but is sensitive to noise and it is difficult to calculate the distances precisely.
- The topological framework only considers places and relations between them. Often, the distances between places are stored. The map is then a graph, in which the nodes corresponds to places and arcs correspond to the paths.

Many techniques use probabilistic representations of the map, in order to handle uncertainty.

There are three main methods of map representations, i.e., free space maps, object maps, and composite maps. These employ the notion of a grid, but permit the resolution of the grid to vary so that it can become finer where more accuracy is needed and more coarse where the map is uniform.

MAP LEARNING

Map-learning cannot be separated from the localization process, and a difficulty arises when errors in localization are incorporated into the map. This problem is commonly referred to as Simultaneous localization and mapping (SLAM).

An important additional problem is to determine whether the robot is in a part of environment already stored or never visited. One way to solve this problem is by using electric beacons.

PATH PLANNING

Path planning is an important issue as it allows a robot to get from point A to point B. Path planning algorithms are measured by their computational complexity. The feasibility of real-time motion planning is dependent on the accuracy of the map (or floorplan), on robot localization and on the number of obstacles. Topologically, the problem of path planning is related to the shortest path problem problem of finding a route between two nodes in a graph.

ROBOT NAVIGATION

Outdoor robots can use GPS in a similar way to automotive navigation systems. Alternative systems can be used with floor plan instead of maps for indoor robots, combined with localization wireless hardware. Electric beacons also have been proposed for cheap robot navigational systems.

MOBILE WHEELED ROBOT FOR MECHATRONIC CONTEST

The mobile robot designed for a line following competition and obstacles avoidance. For this type of mobile robot tasks looks for energetic and informational autonomy. Informational autonomy refers to the ability of the robot to perform the task for which it was programmed and can make decisions in extreme situations [1]. Line following contests usually consists in a sinuous black path, 15mm or 19mm wide on a white background with arcs with a minimum radius of 100mm and intersections at right angles. In some cases, there may route markings adjacent track line perpendicular to it, which are placed at the beginning of each arc. These marks are used for robots to adapt the speed while cornering.

ROBOT TASK DEFINITION

The robot, which size should not exceed 200mm x 200mm x 200mm, must be able to travel independently in short time, two types of routes: a speed one and an obstacle one. Speed route is a route with simple, straight lines and arcs of minimum radius of 100mm, without junctions or line breaks. Obstacle courses consists of straight lines, arcs of minimum radius of 100mm and the obstacles that intersection at right angle turns at right angles, consecutive series of turns at right angles to form a step ladder with 30mm, roundabouts, line stopping and resuming it after 50mm in any direction and line break and replace it with an auxiliary wall.

Line follower robot mechatronic structure

The proposed robot is a wheeled mobile robot with two differential type wheels and a rear one for support because it is the most effective type of robot for this task. The chassis is constructed of steel sheet to ensure the necessary rigidity at low cost. On this chassis are fixed motors and wheels, support wheel, electronic circuits, battery and front panel sensor.

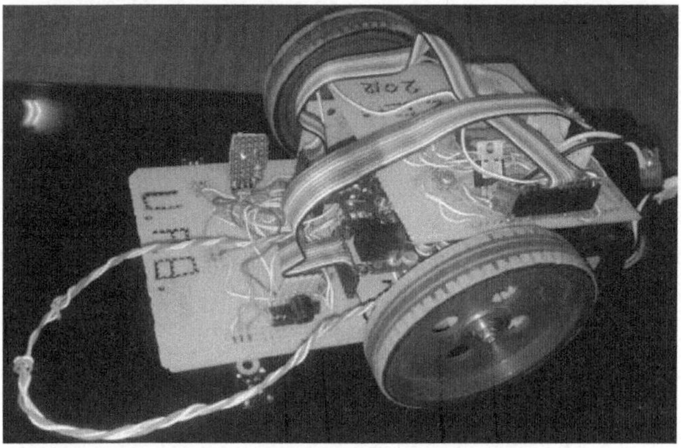

Fig. Mobile robot for line following competitions: 3D – CAD (left) and the final product (right)

All components are mounted as to provide a center of gravity located low down and close to the axis wheels. 1331T012SR motors used are from MINIMOTOR, coupled with planetary gear transmission ratio of 14:1 and incremental rotary transducers.

In terms of control electronics, microcontroller Atmel ATmega 2560 used is mounted on a development board Arduino Mega. The development board was chosen because it contains all the necessary electronics for the operation and programming of microcontroller and Arduino programming environment contains many powerful libraries that simplify complex programming tasks [3].

For motor drives was used a L298 chip from STMicroelectronics which serves dual H-bridge. The battery used is composed of 3 cell lithium polymer capacity 2Ah. It provides a nominal voltage of 11.1 V. For line detection, the robot is equipped with reflective sensors [2]. The sensors used are manufactured by Parallax QTI type.

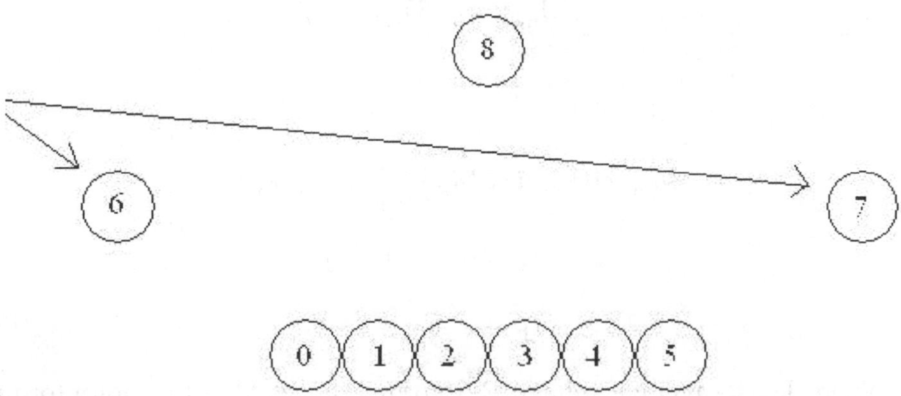

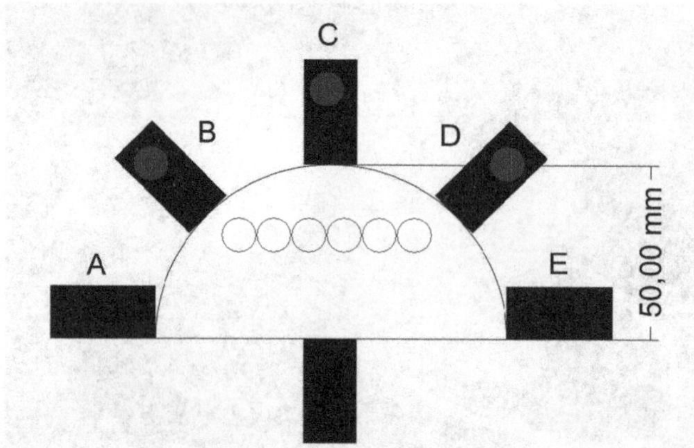

Fig. Sensors placement and an example for one kind of an obstacle

The controllers are implemented on the microcontroller. It receives analog signals from the line detection sensors and, using analog-digital convertor (ADC), the values are converted into digital values which then will be processed. The motor rotary encoder produces two digital signals that are also processed by the microcontroller. The results of processing are two type pulse width signals and two digital signals for reversal. These four signals are fed into an integrated circuit L298 dual H-bridge type which supplies the motors. The general block diagram has another block regarding wall obstacle strategy, in fact reading of the specialised sensor.

Mathematical model

Starting from the position of gravity centre and the reaction between wheels and ground terrain, some calculations were made in order to know the values for actuation force at the ground level function of the nominal battery voltage Ubat and duty cycle fu. Those values are very important for the command strategy.

$$U_{med} \ @ \ f_u \ \text{ð} U_{bat}$$

$$I \ @ \ \frac{U_{med}}{R}$$

$$M_m \ @ \left(\left(k_m \ \text{ð}i\right) 0 \ M_R\right) \text{ó} \frac{1}{red} \text{ð} k_{red} \ \text{ð} k_t$$

$$M_a \ @ \ M_m \ 0 \ M_f$$

$$F_a \ @ \ \frac{M_a}{r} \ \text{ð}2$$

Where: U_{med} - average voltage, R- rotor resistance, M_m - the motor torque at the axis of the actuating wheel, i – motor current, M_R – friction torque at the motor bearings level, red – gear transmission coefficient, k_t - 1:1 transmission efficiency, k_{red}- gear efficiency, Ma – actuating torque for actuating wheels, Mf is friction torque, r is the wheel radius, Fa – actuating force for actuating wheels.

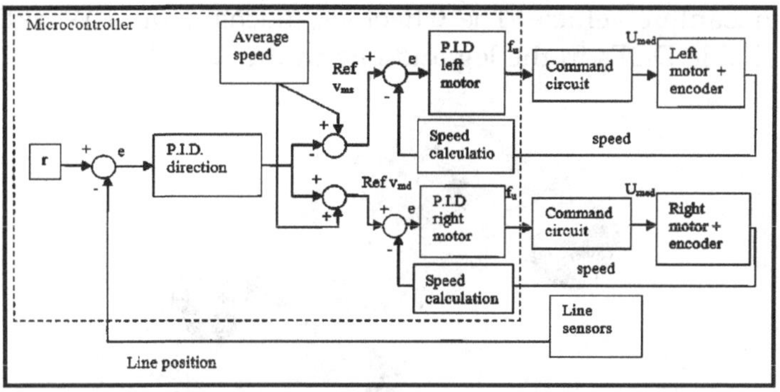

Fig. Control block diagram for the speed contest

By developing a mathematical model of the robot implemented in Matlab/ Simulink environment, the device behavior was observed for various speeds, analyzing the power consumption and operating force developed. In this way were achieved an optimization of control.

ROBOT MOTION BY SIMULTANEOUSLY WHEEL AND LEG PROPULSION

WorkPartner, is a futuristic service robot designed to be used mainly in urban outdoor environment. Mobility is based on a hybrid locomotion system, which combines benefits of both legged and wheeled locomotion to provide good terrain negotiating capability and large velocity range at the same time. The mobile platform, called Hybtor, on which the manipulation and tooling system is built. The WorkPartner project and its mechatronic design have been reported in three previous CLAWAR conferences (1), (2), (3). The purpose of present paper is to continue the series by introducing development, which is done for its hybrid locomotion and motion control system.

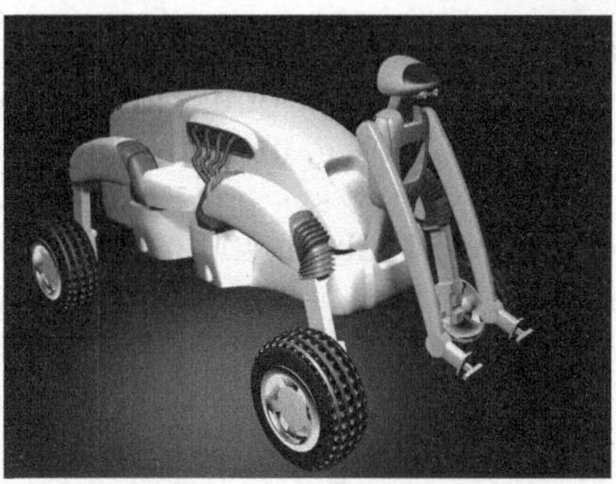

Fig. Design picture of the WorkPartner robot

Similar locomotion system have been reported earlier e.g. by Hirose (4) and Zanthic Technologies Inc (5). These systems are not, however, so complete as to their controllable DOFs per leg as in this case. Either the wheels are not actively controllable like in Hirose machine or the legs have less than three DOFs like in Zanthic vehicle. The wheel-leg mechanism of WorkPartner has three controllable DOFs in the leg and one in the wheel.

Fig. Hybtor platform

MECHANICAL AND SOFTWARE ARCHITECTURE STRUCTURE

Platform

The platform has an active body joint and four legs equipped with wheels. Each leg has three actively controlled joints and the wheel is also actively controlled. The weight is about 200 kg, including all mechanical components and the components of the energy system, the actuating system and the computing system. The payload is about 40 kg, which is mainly taken by the two-hand manipulator system. The actuation system is fully electrical and the power system a hybrid one with batteries and 3 kW combustion engine. The locomotion system allows motion by legs only, by legs and wheels powered at the same time or by wheels only. The robot can obtain 7 km/hour maximum speed when moving by wheels on a hard flat ground.

Manipulator

WorkPartner robot has a two-arm manipulator system illustrated in Fig. 3. It consists of a two degree of freedom body, two 3 degree of freedom arms and a two-degree of freedom camera head. The manipulator's body is jointed to the platform with two joints, which allow orientation to horizontal and vertical directions. With the manipulator installed into the front of the body WorkPartner looks like a centaur, so instead of humanoid it may be called centaur.

Fig. WorkPartner with the manipulator

With two arms manipulator the robot is to be able to operate with large objects, the sizes of which exceed the arms grippers' size. This is important in many service tasks. With two hands the manipulator can handle objects up to 10 kg at 1 meter range. Motion of the manipulator and platform is coordinated by the overall motion control system.

General software architecture

The main software architecture of the robot. A systematic way to develop the software is considered very important in the project because of its large scale and number of people included. The robot will also evolve through several generations, which causes special needs to modular design and documentation of the software. Software planning and designing is done using UML description and the program code is written using object-oriented methods.

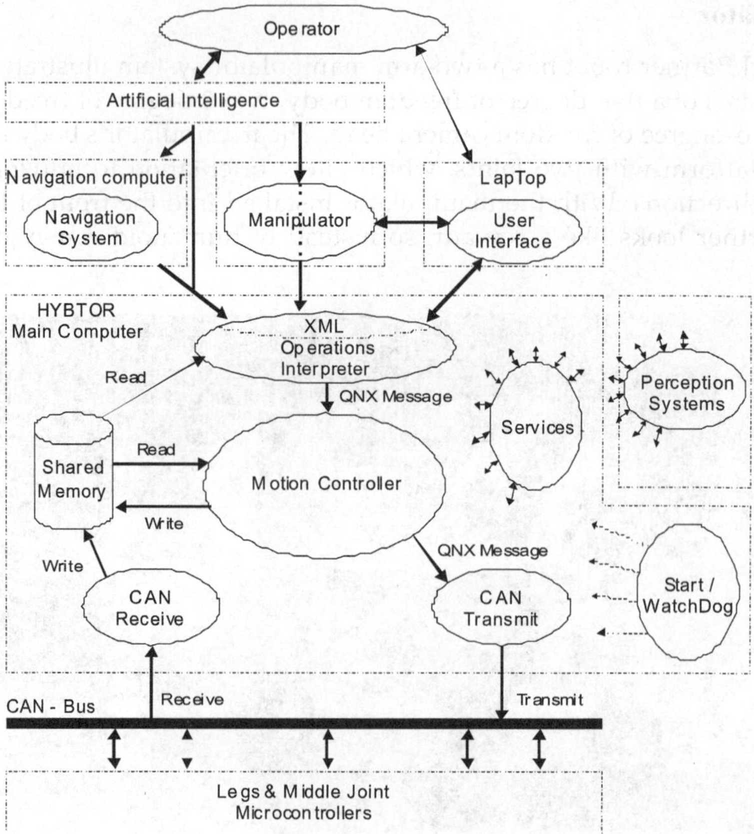

Fig. Software architecture of the WorkPartner robot

The software is divided into hierarchical levels, which communicate with each other using a special WorkPartner commanding language, which will be constructed by using the notations of the XML.

The operator can use the robot trough two interfaces. The conventional interface, so called test interface, is built with a laptop and joystick. The advanced interface is a cognitive interface which utilizes different communication tasks on AI interpretation software. Working tasks are taken care by the manipulator system and transferring tasks by the navigation system. The Hybtor platform produces mobility services for the other tasks. Services of the Perception Systems are available throughout the architecture. A standard user interface can be used parallel to the AI level more advanced in future.

SOFTWARE OF THE HYBTOR PLATFORM

The software of the Hybtor platform is divided into separate parallel processes. Start/Watchdog activates all the other tasks in a predetermined order. Then it stays awake and guards the states of all the other tasks.

Operations Interpreter acts as a interface between the platform and the upper hierarchy levels. Motion Controller takes care of all the calculations concerning the motion planning of the platform. Shared Memory contains all the measured data from the legs and the middle joint. Can Receive reads the measurements from the CAN-bus and writes them into the shared memory. Can Transmit sends all the controlling commands to the individual leg controllers or to the middle joint controller.

Simultaneous wheel and leg locomotion

The locomotion system allows motion by legs only, by legs and wheels powered at the same time or by wheels only. Hybrid locomotion means combining the wheeled and legged locomotion modes so that the propulsive force is generated by the wheel and the leg joints simultaneously. Hybrid locomotion could be called also 'rolking' (rolling-walking). A term close to this one (roller-walker) has been used previously by Hirose (4), but his rolking robot differs from WorkPartner in that its wheels are not powered. Rolking in this case resembles skiing, but instead of skis wheels are used (however, skis are not active devices like wheels in this case).

Rolking resembles a normal walking sequence. When a leg is in the supporting state the propulsive force is generated by distributing the moments between the leg joints and the wheel joint. How the distribution is done depends on the terrain properties and the speed of the platform. When the leg is in the transferring phase it is not lifted in the air, but unloaded and moved along the ground by touching it all the time and applying a slight forward moment to the wheel at the same time. All the joints of the leg, hip, thigh, knee and wheel are thus controlled actively all the time. Thus, for example, COG can be controlled during locomotion which is important when legs are moved.

In the transferring phase it is possible to 'feel' the shapes of the ground and detect the obstacles by measuring the actuator currents and the joint angles (the leg is in a force control mode). The robot can then move on an uneven terrain by "probing" by the aid of feet. On a highly soft terrain (e.g. on sand or snow), where wheeled locomotion is difficult or impossible, it has been experimentally observed that rolking motion can improve mobility considerably.

Other benefits of the rolking mode compared to normal walking are better speed, stability and weight distribution of the platform. The leg can be moved to supporting phase instantly if needed, which improves the reaction responses. Speed is improved because there is no time wasted when lifting or lowering the leg in the walking cycle. Stability will not be easily lost and the weight distribution is more equally divided because the transferring leg supports itself when moving. Standard gait algorithms can be used. When the gait algorithm command a transferring leg to the supporting phase, it can be done instantly because the leg is already on the ground. This is very effective especially when free gait algorithms are used, which seems to be the natural choice in this case.

The main disadvantage of the rolking mode, if compared to the walking mode, is that the legs can be moved to the same direction as the wheels are rolling. The motion direction must be thus controlled like in the wheeled mode. In the case of Hybtor, steering is done by using the articulated body. A minor disadvantage is due to the fact that legs are always contacting to the ground, which may cause problems in some applications (e.g. few foot places available on sensitive terrain). On the other hand, changing between the different locomotion modes is very simple and in fact they are all controlled by the same program using only different parameters.

Overrunning an obstacle by the aid rolking locomotion mode

In rolking, the transfer leg supports lightly the body and the wheel rotates actively by the velocity control. When the wheel hits the obstacle and is not able to go forward, the control algorithm starts to lighten the leg. Thigh and knee currents change according to the force reference. Wheel current changes according to the velocity reference.

Overrunning an obstacle is based on simultaneous lightening of the leg and active rolling of the wheel. When the wheel is on the obstacle, lightening of the leg stops but the wheel continues rotation forward pulling the leg over the obstacle smoothly. Tests show that it is relative easy to find out a rough form of the obstacle, which we think can be utilised in several way in the future development of robot.

When overrunning of an obstacle, stability of the robot must be maintained. Center of gravity has to be transferred towards support legs. Climbing up stairs has been successfully demonstrated by the aid of the dynamic simulator of the robot. Climbing tests with the real robot are presently carried out.

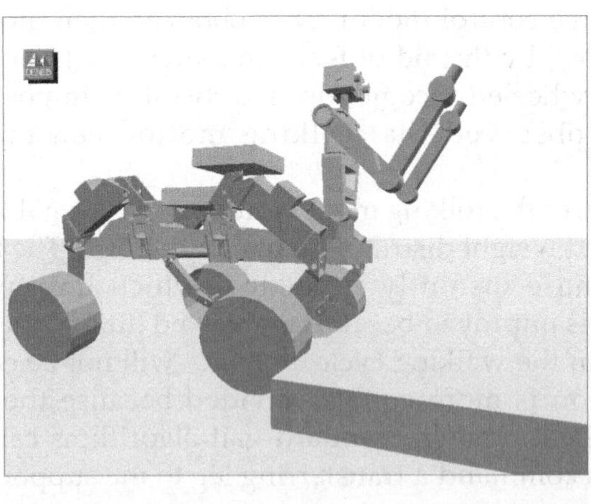

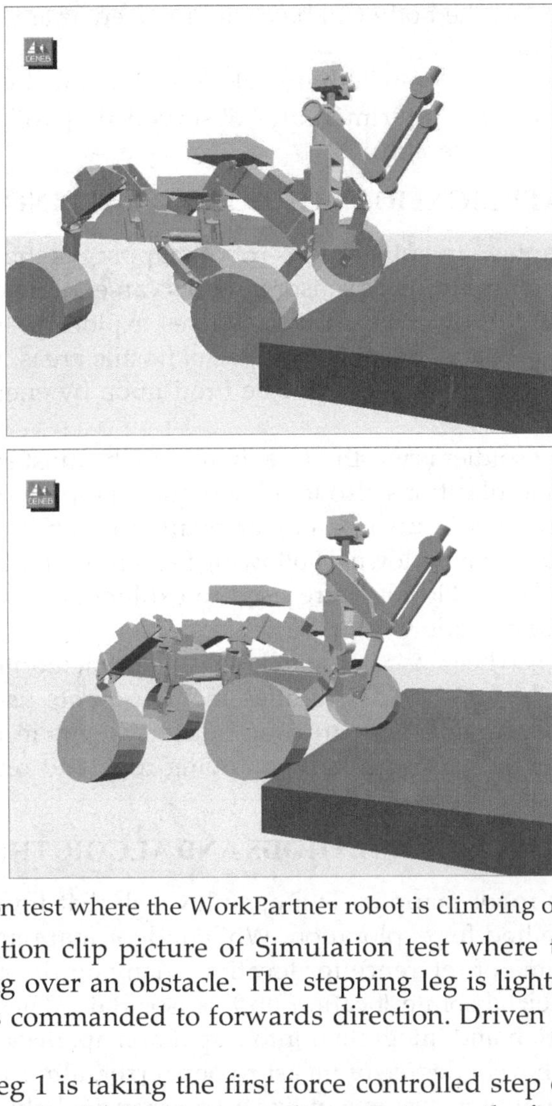

Fig. Simulation test where the WorkPartner robot is climbing over an obstacle

One animation clip picture of Simulation test where the WorkPartner robot is climbing over an obstacle. The stepping leg is lightened and a force action of 85N is commanded to forwards direction. Driven wheel helps also climbing.

Left front leg 1 is taking the first force controlled step over the obstacle and the other front leg 2 takes the second step. Between the first and the second step weight shifting is needed by changing supporting rear legs configuration and by using hip inclination.

Conclusions

The rolking (rolling walking) locomotion mode seems to be a natural and effective way of moving for hybrid wheeled-legged machines. It combines the good features of both legs and wheels in difficult terrain conditions. Terrain unevenness can be detected and relative big obstacles can be negotiated easily. In addition, stability of the machine is easily maintained because reflexive leg motion to support the body can be easily realized at any moment during the leg transfer state.

On a highly soft terrain, where wheeled locomotion is difficult or impossible, it has been experimentally observed that rolking motion can improve mobility considerably.

APPLICATIONS OF ROBOT MAPPING

The key advantage to autonomous robot mapping is the use of the robot itself. Along with a multitude of sensors, robots can explore dangerous areas that humans would not want to risk their lives exploring. For instance, the military has used robots to map out and patrol hostile areas. If a soldier were to map the same area, he or she could be fired upon by enemy soldiers and killed.

The life of that soldier is worth more than even the most expensive robots. Some commercial applications also involve robots mapping dangerous areas. Robots have been used in nuclear power plants to map out high-radiation areas following nuclear meltdowns. Following September 11, 2001, robots from the University of South Florida were used to explore and map out unstable tunnels in the World Trade Center.

Robots have also been used to search volcanoes, meteorites in Antarctica, and map the sea bed. Other possibilities include robots used for mapping and other tasks. Currently there are warehouse systems in design that will map a warehouse, then use the same mapping robots to retrieve inventory that the workers specify.

AUTONOMOUS MAPPING METHODS AND ALGORITHMS

Autonomous robot mapping is a complex task that involves a sense of location and a method for exploration. While indoor mapping systems must find a relative frame of reference for location, many outdoor systems utilize GPS to record their absolute location and sensor data. These systems then take this information and integrate it into a spatial map. Because both robotic sensors and GPS have a degree of measurement error, almost all state-of-the-art autonomous mapping systems utilize a Bayesian probability algorithm to infer information about objects and obstacles from the data. Many systems contain algorithms modified for their specific sensor types and sensor performance.

Robot mapping systems must also create an algorithm for exploring areas. These algorithms can range from basic to complex. The most basic algorithm is to start at a point, go straight until an obstacle is reached, and then return to that point and repeat. More complex algorithms may involve using probabilities to estimate where walls or obstacles are. The most efficient way to explore is the use of multiple robots. This allows the mapping to be divided and finished quicker. However, the use of multiple robots adds another layer of design to the system. Almost all multiple robots mapping systems require the robots to have some sense of relative location to the other robots. The system must also deal with integrating data from multiple robots which can lead to uncertainty when separate robot maps are pieced together. The Distributed Multi-robot Exploration and Mapping Group at University of Washington has designed an innovative system in which multiple robots will map out an area while actively searching for another robot to exchange sensor data with. After this exchange, the robots physically verify their integrated maps for increased accuracy then divide remaining areas for faster exploration.

Typical Mapping Robot Sensors and Parts

A typical robot used for autonomous mapping will require parts for location, movement, and recording data. Most outdoor systems use GPS receivers for locating the robot, while indoor systems use a combination of sensors and reference points. Kiva Systems Inc. has designed an indoor warehouse mapping system that uses barcodes glued to the floor. Indoor robots typically use wheels and motor to move around while a sturdier outdoor mapping robot may need treads such as those found on tanks to navigate bumpy terrain. All mapping robots have a distance sensor such as infrared or sonar and touch. Specialized mapping robots have sensors for their specific application. For example, robots mapping nuclear power plants would have radiation detectors. When multiple robots are used in mapping, a wireless communications medium such as WiFi must be added to each robot.

4

Mobile Robot Kinematics

ROBOT KINEMATICS

Robot kinematics applies geometry to the study of the movement of multi-degree of freedom kinematic chains that form the structure of robotic systems.

The emphasis on geometry means that the links of the robot are modeled as rigid bodies and its joints are assumed to provide pure rotation or translation.

Robot kinematics studies the relationship between the dimensions and connectivity of kinematic chains and the position, velocity and acceleration of each of the links in the robotic system, in order to plan and control movement and to compute actuator forces and torques. The relationship between mass and inertia properties, motion, and the associated forces and torques is studied as part of robot dynamics.

KINEMATIC EQUATIONS

A fundamental tool in robot kinematics is the kinematics equations of the kinematic chains that form the robot. These non-linear equations are used to map the joint parameters to the configuration of the robot system. Kinematics equations are also used in biomechanics of the skeleton and computer animation of articulated characters.

Forward kinematics uses the kinematic equations of a robot to compute the position of the end-effector from specified values for the joint parameters. The reverse process that computes the joint parameters that achieve a specified position of the end-effector is known as inverse kinematics. The dimensions of the robot and its kinematics equations define the volume of space reachable by the robot, known as its workspace.

There are two broad classes of robots and associated kinematics equations serial manipulators and parallel manipulators. Other types of systems with specialized kinematics equations are air, land, and submersible mobile robots, hyper-redundant, or snake, robots and humanoid robots.

Forward kinematics

Forward kinematics specifies the joint parameters and computes the configuration of the chain. For serial manipulators this is achieved by direct substitution of the joint parameters into the forward kinematics equations for the serial chain. For parallel manipulators substitution of the joint parameters into the kinematics equations requires solution of the a set of polynomial constraints to determine the set of possible end-effector locations. In case of a Stewart platform there are 40 configurations associated with a specific set of joint parameters.

Inverse kinematics

Inverse kinematics specifies the end-effector location and computes the associated joint angles. For serial manipulators this requires solution of a set of polynomials obtained from the kinematics equations and yields multiple configurations for the chain. The case of a general 6R serial manipulator (a serial chain with six revolute joints) yields sixteen different inverse kinematics solutions, which are solutions of a sixteenth degree polynomial. For parallel manipulators, the specification of the end-effector location simplifies the kinematics equations, which yields formulas for the joint parameters.

ROBOT JACOBIAN

The time derivative of the kinematics equations yields the Jacobian of the robot, which relates the joint rates to the linear and angular velocity of the end-effector. The principle of virtual work shows that the Jacobian also provides a relationship between joint torques and the resultant force and torque applied by the end-effector. Singular configurations of the robot are identified by studying its Jacobian.

Velocity kinematics

The robot Jacobian results in a set of linear equations that relate the joint rates to the six-vector formed from the angular and linear velocity of the end-effector, known as atwist. Specifying the joint rates yields the end-effector twist directly.

The inverse velocity problem seeks the joint rates that provide a specified end-effector twist. This is solved by inverting the Jacobian matrix. It can happen that the robot is in a configuration where the Jacobian does not have an inverse. These are termed singular configurations of the robot.

Static force analysis

The principle of virtual work yields a set of linear equations that relate the resultant force-torque six vector, called a wrench, that acts on the end-effector to the joint torques of the robot. If the end-effector wrench is known, then a direct calculation yields the joint torques. The inverse statics problem seeks the end-effector wrench associated with a given set of joint torques, and requires the inverse of the Jacobian matrix. As in the case of inverse velocity analysis, at singular configurations this problem cannot be solved. However, near singularities small actuator torques result in a large end-effector wrench. Thus near singularity configurations robots have large mechanical advantage.

FIELDS OF STUDY

Robot kinematics also deals with motion planning, singularity avoidance, redundancy, collision avoidance, as well as the kinematic synthesis of robots.

KINEMATICS MODEL OF NONHOLONOMIC WHEELED MOBILE ROBOTS

The robotized technologies are very quickly spreading for domestic, service and entertainment needs. A big number of scientific investigations and scientific activities form a new scientific field during the recent years devoted to mutual interaction among robots and the human being. This is an interdisciplinary scientific field that covers robotics, computer sciences, and the science of knowledge, physiology and sociology.

Robots are going very soon to assist the human being on a wide range of problems, which are not attractive, they are dangerous, not well paid or boring to humans. Robots assistants are going to work in the future as patient sitters, as security guards, as rescuers and fire rescuers, in surgery and rehabilitation, in domestics and in offices, in mining, in building as well as in stores and museums.

In order to work together with and to assist and to interact with people the new robot generation must posses a mechanical structure that is suitable for this partnership in the human not organised and unknown surrounding environment. The robot's successful invasion in this environment depends on the development of competent and practical systems that can be very well trusted, which are reliable and simple for usage [1,2,3,4]. Wheeled mobile robots (WMR) are long ago invented in areas, where they interact with humans as service robots for remote book reading in a library or for serving tea and meals, human following and guiding robots, entertainment and cleaning robots.

The compatible with people robots must integrate mobility and manipulation. Mobile manipulator systems hold promise in many industrials and service applications including manipulations, assembly, inspection and work in hazardous environments. The integration of a manipulator and a mobile robot base places special demands on the vehicle's mechanical system.

The objective of the present paper is to evaluate the possible solutions of mobile robot bases and to build a kinematics model of WMR suitable for mobile manipulation tasks.

HOLONOMIC AND NONHOLONOMIC WMR

The wheeled mobile robots (WMR) are divided into two basic types - holonomic and nonholonomic. Theoretically, the holonomic mechanical systems comprise links, that impose restrictions on the limb velocities, and after integration these restrictions can be reduced to restrictions only on the limb locations.

When WMR do not impose restrictions on the motion velocities in the 2D (planar) solution they are called holonomic. The holonomic WMR possess maximal number of degrees of freedom in the 2D (planar) solution h=3. In the field of the mobile robots the term holonomic is used as an abstract term for WMR with three degrees of freedom. Thus, every WMR with three degrees of freedom in plane is called a holonomic one. Various mechanisms are used as universal or omni wheels, orthogonal or ball wheels in order to achieve a holonomic motion. [5]. ?he holonomic WMR allow easier motion planning in plane. The nonholonomic mechanical systems comprise links restricting the system velocities; thus these restrictions cannot be integrated. In this way the nonholonomic WMR impose restrictions on the velocities of the motions in plane.

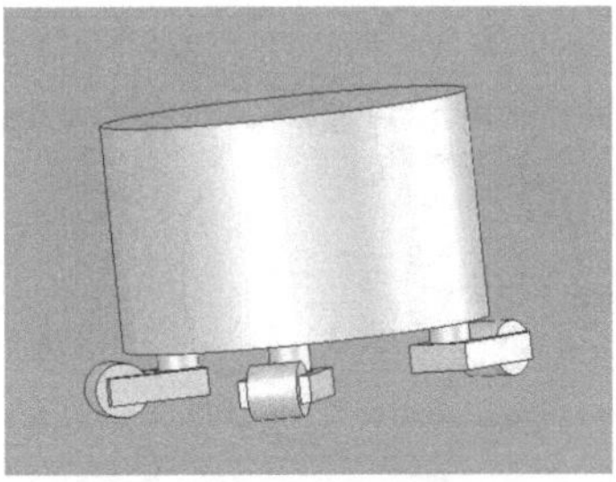

Fig. Holonomic WMR

Due to this reason the nonholonomic WMR possess less than three degrees of freedom in plane h < 3. They are simpler in construction and thus cheaper, with less controllable axes and ensure the necessary mobility in plane. Due to this reason a kinematics model of nonholonomic WMR with two degrees of freedom h=2 are derived in the present work.

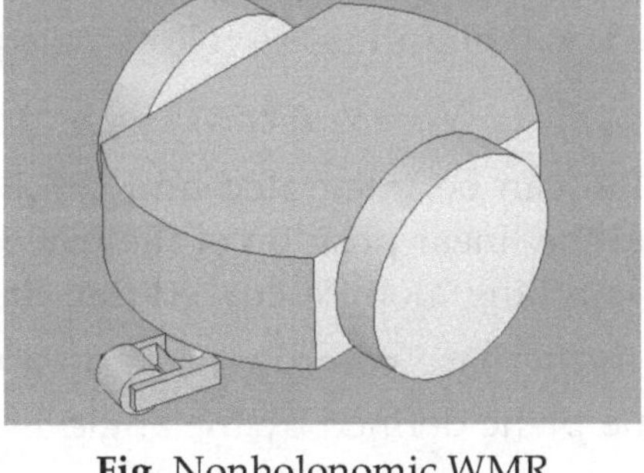

Fig. Nonholonomic WMR

A generalised model of a nonholonomic WMR with h=2 is presented. It includes two symmetrically allocated driving wheels with radii r. The nonholonomic WMR include a various number universal wheels for keeping up the balance in plane.

The investigated WMR possess a single universal wheel. This wheel is not a driving one and it is not included in the kinematics model.. In the robot centre P is connected a local co-ordinate system PX_1Y_1, where X_1 is along the axis of symmetry, and Y_1 is along the axis of the driving wheels. The angle between the axis X_1 and the axis X of the immovable co-ordinate system OXY is denoted with i.

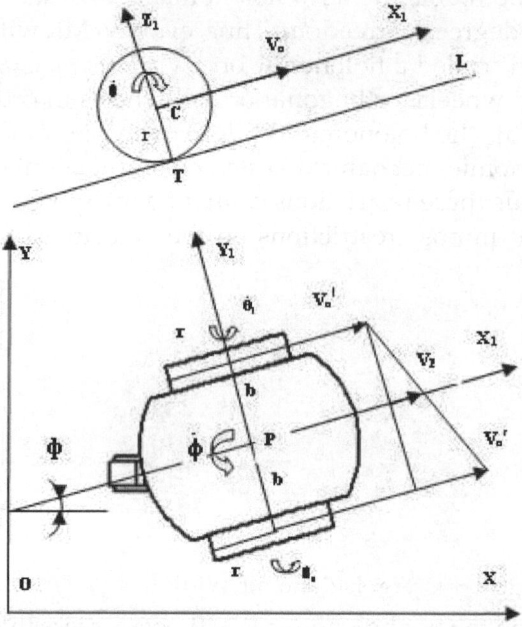

Fig. A general model of non-holonomic WMR.

The distance between the driving wheels along the axis Y_1 is 2b, and the angular velocities of the left and the right wheel are respectively t_l,#t_r.

When one wheel is rolling on a straight line without slipping with angular velocity t, its centre is moving with velocity Vc. The velocity of the oscillate point T with the plane L is 0 and thus the following equation is fulfilled:

$$V_T \;@\; V_c \;0\; \mathring{t}r \;@\; 0$$

The upper equation can be integrated and can be presented as a link among the angular and the linear position of the wheel: l - ?r = 0.

When the wheel is rolling along a curved line the linear velocity of its centre $V_c \;@[\mathring{X}_c;\mathring{Y}_c]^T$ in the base co-ordinate system OXY depends on the wheel orientation in the plane defined by the angle i

$$\begin{vmatrix} \dot{X}_c - \dot{\varphi} r \cos\theta = 0 \\ \dot{Y}_c - \dot{\varphi} r \sin\theta = 0 \end{vmatrix}$$

The upper equations can not be integrated in order to define relations only between the wheel positions. In the plane motion on the wheel velocities are imposed restrictions, thus the mobile devices from the type are called nonholonomic WMR.

DIRECT AND REVERSE KINEMATICS TASK OF NONHOONOMIC WMR

The nonholonomic mobile devices include two co-axial driving wheels, the velocities of their centres V_c^r H V_c^l are co-linear with the axis X_1 of the local co-ordinate system PX_1Y_1. The velocity Vp of the centre P of the mobile platform is also co-linear with the axis X_1. The plan of velocities of WMR in the plane OXY. The following equations can be derived:

$$V_p = 1/2(V_c^r + V_c^l)$$

$$2b\dot{\theta} = V_c^r - V_c^l$$

If we derive the upper equations along the axes of the base co-ordinate system OXY, where the velocity of the centre V_p is presented by the co-ordinates $V_p^0 = [\dot{X}_p; \dot{Y}_p]^T$, and the velocities V_c^r H V_c^l of the right and the left wheel is defined with the help of (2), thus equations are derived defining the kinematics of WMR in the base co-ordinate system

$$\begin{vmatrix} \dot{X}_p = \frac{r}{2}\cos\theta\dot{\varphi}_r + \frac{r}{2}\cos\theta\dot{\varphi}_l \\ \dot{Y}_p = \frac{r}{2}\sin\theta\dot{\varphi}_r + \frac{r}{2}\sin\theta\dot{\varphi}_l \\ \dot{\theta} = \frac{r}{2b}\dot{\varphi}_r - \frac{r}{2b}\dot{\varphi}_l \end{vmatrix}$$

If the velocity V_p of the centre of the mobile platform and its velocity of rotation i we combine in the vector:

$$\dot{X} = [\dot{X}_p; \dot{Y}_p; \dot{\theta}]^T$$

and the velocities of the driving wheels $t_l, \#t_r$ we combine in the vector

$$\dot{\varphi} = [\dot{\varphi}_r; \dot{\varphi}_l]^T$$

then the direct task of the kinematics of WMR is presented by the vector equation,

$$\dot{X} = S\dot{\varphi}$$

where,

$$S = \begin{bmatrix} \frac{r}{2}\cos\varphi; & \frac{r}{2}\cos\varphi \\ \frac{r}{2}\sin\varphi; & \frac{r}{2}\sin\varphi \\ \frac{r}{2b}; & -\frac{r}{2b} \end{bmatrix}$$

The direct task allows for each position of WMR definition of (3x1) the vector of velocities of the centre P from (2x1) the vector of velocities of the driving wheels.

It is necessary to solve the reverse task of the kinematics in order to plan the robot motion and the robot control. In order to find a solution for the reverse task of the kinematics of WMR, the pseudo inverse matrix S+ of can be used, because matrix is not a quadratic one.

$$\dot{\theta} = S \cdot \dot{X}$$

It is necessary to use an additional restriction on parameters defined by the non-holonomic links of WMR, because in this case the number of the input parameters is bigger than the number of the output parameters.

$$A\dot{X} = 0$$

Matrixes S+ and A can be easily defined by using the equalities presented in the form below:

$$\begin{cases} V_p + b\dot{\varphi} = V_c^r \\ V_p - b\dot{\varphi} = V_c^l \end{cases}$$

when we define the upper equations along the axes of the local co-ordinate system PX_1Y_1. The velocity of the centre $V_p = [\dot{X}_p; \dot{Y}_p]^T$ is defined along the axes X_1 and Y_1 by means of consideration of their angle of rotation ?. In this way equations along axis X1 present:

$$\begin{cases} \dot{X}_p\cos\varphi + \dot{Y}_p\sin\varphi + b\dot{\varphi} = r\dot{\theta}_r \\ \dot{X}_p\cos\varphi + \dot{Y}_p\sin\varphi - b\dot{\varphi} = r\dot{\theta}_l \end{cases}$$

and along axis Y1 present:

$$-\dot{X}_p\sin\varphi + \dot{Y}_p\cos\varphi = 0$$

Equation express the reverse task in kinematics, from where for the matrix S+ we can derive:

$$S^{+} = \frac{1}{r}\begin{bmatrix} \cos\varphi; & \sin\varphi; & b \\ \cos\varphi; & \sin\varphi; & -b \end{bmatrix}$$

Equality expresses the equation of the links between the velocities of the nonholohomic WMR including the matrix:

$$A\ @\ \tilde{A}0 \sin m; \cos m; 0$$

As it can be clearly seen by the developed study on the three independent parameters that define uniquely the positioning of WMR in the plane n=3, that is imposed one nonholonomic restriction m=1, from where the degrees of freedom of the nonholonomic WMR in plane are defined as two h=n-m=2.

Conclusion

The developed kinematics model of the nonholonomic WMR can be used for control creation in mobile tasks or in mobile manipulation tasks. Control can be based on the direct kinematics task by means of using equation. In this case the equilibrium of the given velocities of the wheels t_1, t_r guarantee motion along a straight line, and the difference between them defines robot rotation. This control mode belongs to a lower level and is more convenient for derivation of on-line tasks of motion.

Control can be build up on the inverse kinematics task by using equation. In this case control models are derived including tracing a path considering the non-holonomic restrictions. These control schemes are more sophisticated, they can consider also the dynamics of WMR, they can include adaptiveness and robustness.

5

Aerial Mobile Robots and Vehicle

AEROBOT

An aerobot is an aerial robot, usually used in the context of an unmanned space probe or unmanned aerial vehicle.

While work has been done since the 1960s on robot "rovers" to explore the Moon and other worlds in the Solar system, such machines have limitations. They tend to be expensive and have limited range, and due to the communications time lags over interplanetary distances, they have to be smart enough to navigate without disabling themselves.

For planets with atmospheres of any substance, however, there is an alternative: an autonomous flying robot, or "aerobot". Most aerobot concepts are based onaerostats, primarily balloons, but occasionally airships. Flying above obstructions in the winds, a balloon could explore large regions of a planet in great detail for relatively low cost. Airplanes for planetary exploration have also been proposed.

BASICS OF BALLOONS

While the notion of sending a balloon to another planet sounds strange at first, balloons have a number of advantages for planetary exploration. They can be made light in weight and are potentially relatively inexpensive. They can cover a great deal of ground, and their view from a height gives them the ability to examine wide swathes of terrain with far more detail than would be available from an orbiting satellite. For exploratory missions, their relative lack of directional control is not a major obstacle as there is generally no need to direct them to a specific location.

Balloon designs for possible planetary missions have involved a few unusual concepts. One is the solar, or infrared (IR) Montgolfiere. This is a hot-air balloon where the envelope is made from a material that traps heat from sunlight, or from heat radiated from a planetary surface. Black is the best color for absorbing heat, but other factors are involved and the material may not necessarily be black.

Solar Montgolfieres have several advantages for planetary exploration, as they can be easier to deploy than a light gas balloon, do not necessarily require a tank of light gas for inflation, and are relatively forgiving of small leaks. They do have the disadvantage that they are only aloft during daylight hours.

The other is a "reversible fluid" balloon. This type of balloon consists of an envelope connected to a reservoir, with the reservoir containing a fluid that is easily vaporized. The balloon can be made to rise by vaporizing the fluid into gas, and can be made to sink by condensing the gas back into fluid. There are a number of different ways of implementing this scheme, but the physical principle is the same in all cases.

A balloon designed for planetary exploration will carry a small gondola containing an instrument payload. The gondola will also carry power, control, and communications subsystems. Due to weight and power supply constraints, the communications subsystem will generally be small and low power, and interplanetary communications will be performed through an orbiting planetary probe acting as a relay.

A solar Montgolfiere will sink at night, and will have a guide rope attached to the bottom of the gondola that will curl up on the ground and anchor the balloon during the darkness hours. The guide rope will be made of low friction materials to keep it from catching or tangling on ground features.

Alternatively, a balloon may carry a thicker instrumented "snake" in place of the gondola and guiderope, combining the functions of the two. This is a convenient scheme for making direct surface measurements.

A balloon could also be anchored to stay in one place to make atmospheric observations. Such a static balloon is known as an "aerostat".

One of the trickier aspects of planetary balloon operations is inserting them into operation. Typically, the balloon enters the planetary atmosphere in an "aeroshell", a heat shield in the shape of a flattened cone. After atmospheric entry, a parachute will extract the balloon assembly from the aeroshell, which falls away. The balloon assembly then deploys and inflates.

Once operational, the aerobot will be largely on its own and will have to conduct its mission autonomously, accepting only general commands over its long link to Earth. The aerobot will have to navigate in three dimensions, acquire and store science data, perform flight control by varying its altitude, and possibly make landings at specific sites to provide close-up investigation.

The Venus Vega balloons

The first, and so far only, planetary balloon mission was performed by the Space Research Institute of Soviet Academy of Sciences in cooperation with the French space agency CNES in 1985. A small balloon, similar in appearance to terrestrial weather balloons, was carried on each of the two

Soviet Vega Venus probes, launched in 1984. The first balloon was inserted into the atmosphere of Venus on 11 June 1985, followed by the second balloon on 15 June 1985.

The first balloon failed after only 56 minutes, but the second operated for a little under two Earth days until its batteries ran down.

The Venus Vega balloons were the idea of Jacques Blamont, chief scientist for CNES and the father of planetary balloon exploration. He energetically promoted the concept and enlisted international support for the small project.

The scientific results of the Venus VEGA probes were modest. More importantly, the clever and simple experiment demonstrated the validity of using balloons for planetary exploration.

The Mars aerobot effort

After the success of the Venus VEGA balloons, Blamont focused on a more ambitious balloon mission to Mars, to be carried on a Soviet space probe. The atmospheric pressure on Mars is about 150 times less than that of Earth. In such a thin atmosphere, a balloon with a volume of 5,000 to 10,000 cubic meters (178,500 to 357,000 cubic feet) could carry a payload of 20 kilograms (44 pounds), while a balloon with a volume of 100,000 cubic meters (3,600,000 cubic feet) could carry 200 kilograms (440 pounds).

The French had already conducted extensive experiments with solar Montgolfieres, performing over 30 flights from the late 1970s into the early 1990s. The Montgolfieres flew at an altitude of 35 kilometers, where the atmosphere was as thin and cold as it would be on Mars, and one spent 69 days aloft, circling the Earth twice.

Early concepts for the Mars balloon featured a "dual balloon" system, with a sealed hydrogen or helium-filled balloon tethered to a solar Montgolfiere. The light-gas balloon was designed to keep the Montgolfiere off the ground at night. During the day, the Sun would heat up the Montgolfiere, causing the balloon assembly to rise.

Eventually, the group decided on a cylindrical sealed helium balloon made of aluminized PET film, and with a volume of 5,500 cubic meters (196,000 cubic feet). The balloon would rise when heated during the day and sink as it cooled at night. Total mass of the balloon assembly was 65 kilograms (143 pounds), with a 15 kilogram (33 pound) gondola and a 13.5 kilogram (30 pound) instrumented guiderope. The balloon was expected to operate for ten days. Unfortunately, although considerable development work was performed on the balloon and its subsystems, Russian financial difficulties pushed the Mars probe out from 1992, then to 1994, and then to 1996. The Mars balloon was dropped from the project due to cost.

JPL AEROBOT EXPERIMENTS

By this time, the Jet Propulsion Laboratory (JPL) of the US National

Aeronautics and Space Administration (NASA) had become interested in the idea of planetary aerobots, and in fact a team under Jim Cutts of JPL had been working on concepts for planetary aerobots for several years, as well as performing experiments to validate aerobot technology. The first such experiments focused on a series of reversible-fluid balloons, under the project name ALICE, for "Altitude Control Experiment". The first such balloon, ALICE 1, flew in 1993, with other flights through ALICE 8 in 1997.

Related work included the characterization of materials for a Venus balloon envelope, and two balloon flights in 1996 to test instrument payloads under the name BARBE, for "Balloon Assisted Radiation Budget Equipment". By 1996, JPL was working on a full-fledged aerobot experiment named PAT, for "Planetary Aerobot Testbed", which was intended to demonstrate a complete planetary aerobot through flights into Earth's atmosphere. PAT concepts envisioned a reversible-fluid balloon with a 10-kilogram payload that would include navigation and camera systems, and eventually would operate under autonomous control. The project turned out to be too ambitious, and was cancelled in 1997.

JPL continued to work on a more focused, low-cost experiments to lead to a Mars aerobot, under the name MABVAP, for "Mars Aerobot Validation Program". MABVAP experiments included drops of balloon systems from hot-air balloons and helicopters to validate the tricky deployment phase of a planetary aerobot mission, and development of envelopes for superpressure balloons with materials and structures suited to a long-duration Mars mission. JPL also provided a set of atmospheric and navigation sensors for the Solo Spirit round-the-world manned balloon flights, both to support the balloon missions and to validate technologies for planetary aerobots.

JPL AEROBOT MISSION CONCEPTS

While these tests and experiments were going on, JPL performed a number of speculative studies for planetary aerobot missions to Mars, Venus, Saturn's moon Titan, and the outer planets.

Mars

JPL's MABVAP technology experiments are intended to lead to an actual Mars aerobot mission, named MABTEX, for "Mars Aerobot Technology Experiment". As its name implies, MABTEX is primarily intended to be an operational technology experiment as a precursor to a more ambitious efforts. MABTEX is currently envisioned as a smallsuperpressure balloon, carried to Mars on a "microprobe" weighing no more than 40 kilograms (88 lb). Once inserted, the operational balloon would have a total mass of no more than 10 kilograms (22 lb) and would remain operational for a week. The small gondola would have navigational and control electronics, along with a stereo imagingsystem, as well as a spectrometer and magnetometer.

Current plans envision a follow-on to MABTEX as a much more sophisticated aerobot named MGA, for "Mars Geoscience Aerobot". Current design concepts for MGA envision a superpressure balloon system very much like that of MABTEX, but much larger. MGA would carry a payload ten times larger than that of MABTEX, and would remain aloft for up to three months, circling Mars more than 25 times and covering over 500,000 kilometres (310,000 mi). The payload would include sophisticated equipment, such as an ultrahigh resolution stereo imager, along with oblique imaging capabilities; a radar sounder to search for subsurface water; an infrared spectroscopy system to search for important minerals; a magnetometer; and weather and atmospheric instruments. MABTEX might be followed in turn by a small solar-powered blimp named MASEPA, for "Mars Solar Electric Propelled Aerobot".

Venus

JPL has also pursued similar studies on Venus aerobots. A Venus Aerobot Technology Experiment (VEBTEX) has been considered as a technology validation experiment, but the focus appears to have been more on full operational missions.

One mission concept, the Venus Aerobot Multisonde (VAMS), envisions an aerobot operating at altitudes above 50 kilometres (31 mi) that would drop surface probes, or "sondes", onto specific surface targets. The balloon would then relay information from the sondes directly to Earth, and would also collect planetary magnetic field data and other information. VAMS would require no fundamentally new technology, and may be appropriate for a NASA low-cost Discovery planetary science mission.

Significant work has been performed on a more ambitious concept, the Venus Geoscience Aerobot (VGA). Designs for the VGA envision a relatively large reversible-fluid balloon, filled with helium and water, that could descend to the surface of Venus to sample surface sites, and then rise again to high altitudes and cool off.

Developing an aerobot that can withstand the high pressures and temperatures (up to 480 degrees Celsius, or almost 900 degrees Fahrenheit) on the surface of Venus, as well as passage through sulfuric acid clouds, will require new technologies. As of 2002, VGA was not expected to be ready until late in the following decade. Prototype balloon envelopes have been fabricated from polybenzoxazole, a polymer that exhibits high strength, resistance to heat, and low leakage for light gases. A gold coating is applied to allow the polymer film to resist corrosion from acid clouds.

Work has also been done on a VGA gondola weighing about 30 kilograms (66 lb). In this design, most instruments are contained in a spherical pressure vessel with an outer shell of titanium and an inner shell of stainless steel. The vessel contains a solid-state camera and other instruments, as well as communications and flight control systems. The vessel is designed to tolerate pressures of up to a hundred atmospheres and maintain internal temperatures below 30 °C (86 °F) even on the surface of Venus. The vessel is set at the bottom of a hexagonal "basket" of solar panels that in turn provide tether connections to the balloon system above, and is surrounded by a ring of pipes acting as a heat exchanger. An S-band communications antenna is mounted on the rim of the basket, and a radar antenna for surface studies extends out of the vessel on a mast.

Titan

Titan, the largest moon of Saturn, is an attractive target for aerobot exploration, as it has a nitrogen atmosphere five times as dense as that of Earth's that contains a smog of organic photochemicals, hiding the moon's surface from view by visual sensors. An aerobot would be able to penetrate this haze to study the moon's mysterious surface and search for complex organic molecules. NASA has outlined a number of different aerobot mission concepts for Titan, under the general name of Titan Biologic Explorer.

One concept, known as the Titan Aerobot Multisite mission, involves a reversible-fluid balloon filled with argon that could descend from high altitude to the surface of the moon, perform measurements, and then rise again to high altitude to perform measurements and move to a different site. Another concept, the Titan Aerobot Singlesite mission, would use a superpressure balloon that would select a single site, vent much of its gas, and then survey that site in detail.

An ingenious variation on this scheme, the Titan Aerover, combines aerobot and rover. This vehicle features a triangular frame that connects three balloons, each about two meters (6.6 ft) in diameter. After entry into Titan's atmosphere, the aerover would float until it found an interesting site, then vent helium to descend to the surface. The three balloons would then serve as floats or wheels as necessary. JPL has built a simple prototype that looks three beachballs on a tubular frame.

No matter what form the Titan Biologic Explorer mission takes, the system would likely require an atomic-powered radioisotope thermoelectric generator module for power. Solar power would not be possible at Saturn's distance and under Titan's smog, and batteries would not give adequate mission endurance. The aerobot would also carry a miniaturized chemical lab to search for complicated organic chemicals.

Jupiter

Finally, aerobots might be used to explore the atmosphere of Jupiter and possibly the other gaseous outer planets. As the atmospheres of these planets are largely composed of hydrogen, and since there is no lighter gas than hydrogen, such an aerobot would have to be a Montgolfiere. As sunlight is weak at such distances, the aerobot would obtain most of its heating from infrared energy radiated by the planet below. A Jupiter aerobot might operate at altitudes where the air pressure ranges from one to ten atmospheres, occasionally dropping lower for detailed studies. It would make atmospheric measurements and return imagery and remote sensing of weather phenomena, such as Jupiter's Great Red Spot. A Jupiter aerobot might also drop sondes deep into the atmosphere and relay their data back to an orbiter until the sondes are destroyed by temperature and pressure.

PLANETARY AIRCRAFT

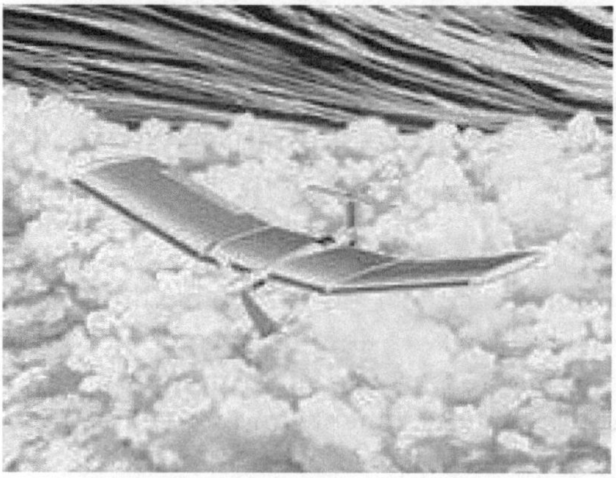

Fig. Artist's conception for a Venus airplane

Winged airplane concepts have been proposed for robotic exploration in the atmosphere of Mars, Venus, and even Jupiter.

The main technical challenges of flying on Mars include:

1. Understanding and modeling the low Reynolds number, high subsonic Mach Number aerodynamics
2. Building appropriate, often unconventional airframe designs and aerostructures
3. Mastering the dynamics of deployment from a descending entry vehicle aeroshell
4. Integrating a non-air-breathing propulsion subsystem into the system.

An aircraft concept, ARES was selected for a detailed design study as one of the four finalists for the 2007 Mars Scout Programopportunity, but was eventually not selected in favor of the Phoenix mission. In the design study, both half-scale and full-scale aircraft were tested under Mars-atmospheric conditions. (See also Mars airplane.)

THE ROBOTICS TRANSFORMATION

Robotics will eventually utterly transform military affairs. Autonomous missiles have a long history and robots are increasingly used for surveillance, auxiliary tasks and armed combat (usually controlled by human operators). From a technical point of view, existing and emerging robot technologies have the potential for rapidly replacing manned platforms with superior autonomous systems in all arenas. Yet, we are still in the early days of the robotics revolution.

One of the main hurdles of the robotics transformation is the lack of integrative standards for robotics. A global standard for plug-and-play robotics would be instrumental for transforming the structure of the industry and dramatically reduce costs thus facilitating the application of robotics to all domains of modern society. Establishing a standard could be the starting point of a dramatic expansion of the robotics industry and market. The power or combination of powers that first succeeds in establishing a plug-and-play standard encompassing a sufficiently large industrial basis would surge far ahead of competitors. Leadership in robotics would quickly translate to tangible economic and military advantages. A plug-and-play standard for robotics is a truly disruptive technology.

To exemplify the application of plug-and-play robotics consider a military task force wishing to deploy robotic scouts for operations in mountainous terrain. Ideally, commanders should be able to ask a provider to quickly assemble customized units from off-the-shelf parts. The integrator would select a six-legged locomotion module from manufacturer A, a communication module from manufacturer B, a power system from manufacturer C, weapons from manufacturer D, sensors from multiple manufactures and a robot operating system from any of a host of providers. As the hardware and software modules are connected to the system they would automatically be integrated in the plug-and-play fashion that we are used to expect from computer components, software and peripherals.

The main work of the provider would be to verify that the pre-tested modules are integrated correctly and perhaps to assist war fighters in configuring the on-board AI for the relevant mission profiles. The cost of the hexapod robot scout would be far lower than present expectations since multiple companies compete for providing parts that are produced in large volumes for a bourgeoning international and mostly civilian market. A plug-and-play standard for robotics would in summary make the robot market of tomorrow operate like the computer market of today in which relevant hardware and software quickly can be assembled for solving the task at hand. Military users would in particular benefit from that components can be replaced in the field and that the same modules can be used in multiple systems.

STANDARDS AS CATALYSERS

The right kind of technical standards may catalyse industrial transformations with paramount consequences for society, life style and economic growth. Major examples in recent times are personal computing and telecommunications. The critical standards can be de facto standards as common in the computer industry or de jure standards as predominantly in telecommunications.

The effect of such disruptive standardization is to replace vertical integration with horizontal integration. Vertical integration means that a single vendor develops the total product that is offered to customers e.g. as Boeing develops aircraft for airline operators. This makes for slow development, expensive products and consumer lock-in. Horizontal integration means that a vendor produces large amounts of similar components that via standardized interfaces can be combined into many different end-user products. End-user products consist of many commoditized components. A personal computer includes e.g. operating system, motherboard, graphics card, keyboard and hard drive.

All these components are produced by many manufacturers with little qualitative differentiation and are easily integrated to a complete system by an integrator company or even a consumer. Horizontal integration leads often to rapid development of inexpensive mass-market products and consumer empowerment. The transition of an industry from vertical integration to horizontal integration is therefore also often a shift from linear growth to exponential growth. A prerequisite for the emergence of a horizontally integrated market is a foundation of components, prototypes and proof-of-concepts. The transition is catalysed by the emergence of de jure or de facto standards that enable sufficiently easy integration of total products that satisfy pent up demands. In this section we study an example of this kind of transition and examine the present status of the robot industry with respect to standardization.

The Personal Computer Transition

Many components and functions of what we now recognize as a personal computer (PC) were in place in 1980. The Commodore PET had a screen, keyboard and a tape recorder for data storage integrated in a single unit [1]. Other vendors had similar products. Yet sales and consumer acceptance were slow. A worldwide total of 724,000 personal computers were shipped by about two dozen companies 1980 [2]. The turning point for the personal computer was when IBM 1981 released an open PC architecture that quickly became the de facto standard. A few years later a host of firms shipped IBM PC clones largely built from third-party components. PC worldwide shipments reached 280 million units in 2009 [3]. Almost half of the homes in highly developed countries have a PC.

The key driver for the rapid PC market penetration was the inadvertent establishment of a de facto standard for hardware and software. Personal computing was before 1981 a jungle of mutually incompatible products where core hardware, peripherals and software largely were specific for each platform. Software applications had to be crafted specifically for each platform. Few applications were hence available. All this changed when the industry leader IBM rushed to launch a PC and cut some corners in the development process by using an open hard-ware architecture and an operating system that also was available to competitors.

Soon many competing products implemented the PC architecture that would become a de facto standard. A new software industry thrived on the expanding market for IBM PC compatible programs and transformed the PC to a ubiquitous and genuinely useful tool for work, communication and entertainment. Standardization also enabled a burgeoning industry of computer peripherals for gaming, communication, printing and many other purposes. In spite of the rapid increase in performance, ease-of-use and applications the cost of owning a PC actually decreased.

From a consumer point of view standardization is noticeable as plug-and-play performance meaning that a new piece of software and hardware works with little or no configuration once it is installed. Because of plug-and-play integration even consumers lacking specialized knowledge and training can build a PC in a few hours – feat that would have been impossible without a standard.

Military forces use PCs for many different tasks and benefit also from the reduced cost of information processing that ultimately is due to the personal computing revolution. To appreciate the power of standards-driven horizontal integration, consider how expensive a modern PC would be if hardware, operating system and applications had been developed by a military contractor according to the vertically integrated business model.

Robotics

The robotics industry of today resembles the computer industry before the 1980s. Multiple providers of vertically integrated systems offer incompatible and expensive problem-specific solutions. Architecture and interfaces are often closely guarded secrets. Any modularization applies typically only to the present provider's portfolio. Customers suffer from systems that are hard to maintain and expand because of technical lock-ins. There are about 30 providers of industry robots but no integrating standard. The need for standardization is equally obvious for military, service and entertainment robots.

Yet there are no fundamental technical reasons for not implementing a generic plug-and-play standard for robotics. Robots are similar to computers in that they include sensors (input devices), actuators (output devices), information storage and processors. Robot sensors overlap partly with computer and game console sensors but are more diverse with e.g. radar, ultrasound and tactile sensors. Robot actuators are also more diverse with complex arms for industry assembly robots, many types of locomotion for mobile robots and weapons for combat robots. Computer sensors and actuators are, however, also increasingly diverse and there is nothing in the broader span of robotic output and input devices that fundamentally prevents the application of plug-and-play technology. Robots must handle real-time operations but real-time operating systems and real-time applications are common also for computers and provide no insurmountable barrier against standardization.

The need for standards in robotics has often been pointed out by experts [4] [5]. So why is the robotics industry locked in a horizontal integration state that obviously hampers growth? One reason may be that the robotics industry, until recently, mainly catered to corporate customers that willingly pay premium prices for industrial robots that replace even more expensive human labour.

The computer industry transition occurred when consumers and small businesses became an important market. In robotics we have just recently seen a rise of consumer service robots (e.g. vacuum cleaners) and entertainment robots that might herald that the robotics business is close to a similar transition point as the computer industry in the beginning of the 1980s. Secondly it might have been an accidental circumstance that an industry leader launched open PC architecture at the right moment. Robotics lacks a dominating industrial player that can enforce a de facto standard.

EMERGING STANDARDS IN ROBOTICS

There are many types of standards that are relevant for robotics. The International Organization for Standardization (ISO) provides e.g. a. safety standard for robots in industrial environments [6] and the NATO Standardization agency supplies standards for UAV airworthiness (STANAG 4671) [7].

This section will, however, focus on representative standardization initiatives that address the key issue of plug-and-play integration of robots. For the sake of brevity we overlook several important standardization and code reuse efforts including the Carnegie Mellon Robot Navigation Toolkit (CARMEN) [8] and the open-source project Player [9]. The seven emerging standards discussed in this section include major open-source projects, corporate initiatives, broad industrial collaborations and initiatives from the defence sector.

The Robot Operating System (ROS) [10] is an open-source framework for mobile robots. ROS emerged 2007 from a Stanford research framework and the driving force is California-based robotics start-up Willow Garage supported by more than twenty partners. ROS includes a peer-to-peer communication system connecting multiple heterogeneous nodes and processes and means for discovering and using services provided by processors, sensors and actuators. An expandable library of packages for relevant functions such as motion tracking, planning, locomotion and grasping ensures a continuous growth of functions.

The main thrust is on UNIX-based systems but a broader support of other platforms is in the pipeline. There seems to be a considerable momentum for ROS in the open-source and academic communities and a recent review of U.S. robotics mentions ROS as the most promising emerging standard [4].

Orca is an open-source framework aiming at promoting re-use of robotics software [11]. It is agnostic of programming language, operating system and architecture. The Internet Communications Engine (ICE) [12] is used as a communications layer, for setting up a registry for service discovery and for managing components. Full Linux and partial Windows and MacOS support are available. The followers are mostly academic but include a few industrial users. The architecture independence and the policy of maximizing cross-platform interoperability at the expense of efficiency makes Orca more of a managed library of re-usable software components and less of a plug-and-play standard.

Sony has since 1997 touted OPEN-R as a standard for entertainment robotics [13]. OPEN-R is a centralized layered architecture where each component carries data describing its function, status and interfaces. Components register automatically with the central CPU thus allowing the robot to discover the present sensor, actuator and software configuration. Sony has since the launch vacillated between openness and protectiveness. This and the basically proprietary nature of OPEN-R have barred any wider industrial adoption.

The Microsoft Robotics Developer Studio (MRDS) [14] facilitates writing programs that concurrently handles multiple sensors and actuators in real-time. MRDS includes a visual programming environment and simulation tools for testing and visualization. Developed programs can be uploaded to robots provided that they run on-board Microsoft operating systems such as Windows CE. Means for hard-ware integration are not provided.

This and the limited choice of operating systems discourage industry-wide adoption. The programming and simulation tools are the main benefits and MRDS are therefore presently mainly used for research and by hobbyists. MRDS could, however, be a forerunner of a Microsoft-based generic robotics platform.

The Object Management Group (OMG) develops enterprise integration standards including the Unified Modelling Language (UML) and the Model-Driven Architecture (MDA). The over 800 members include, Hewlett-Packard, IBM, Lockheed Martin, Microsoft, Nortrop Grumman, THALES and Unisys. The Robotics Domain Task Force (DTF) of OMG aims at extending OMG standards to robotics including both software and hardware and to increase interoperability in robotics by collaborating with other standardization organizations. The OMG program is supported by leading Asian players such as the Japan Robot Association (JARA) and The (Japanese) National Institute of Advanced Industrial Science and Technology (AIST).

AIST has developed an open implementation of the OMG specifications [15]. The OMG Robotic Technology Component Specification describes a component model consisting of a platform-independent part and several platform-specific models. The platform-specific models include a model where components communicate directly and a distributed model employing CORBA-based middleware. DTF develops presently a standard for dynamic deployment and configuration of robotic technology components that will bring it even closer to the plug-and-play objective.

Space Plug-and-Play Avionics (SPA) is a set of standards primarily intended for rapid integration of satellites [16]. SPA is developed by the U.S. Air Force Research Laboratory and has been proposed for standardization to the American Institute for Aeronautics and Astronautics (AIAA). Each component in an SPA system includes a data sheet (xTEDS) containing a description of all commands that can be handled and all messages that can be produced by the component.

The xTEDS is essentially a machine-readable manual for using the component. The components are connected with data transport means selected from a standardized set of specifications that includes the Universal Serial Bus (USB). A new component (e.g. a radar sensor) that is connected to a SPA system is automatically recognized by the other modules. An on-board processor could read the xTEDS of the new component, send commands to the sensor and subscribe to data generated by the sensor. The ontology supporting interpretation of component data sheets is presently limited to satellites but could be generalized to robotics.

The SPA specifications are freely available with the exception of the ontology (Common Data Dictionary) that is needed for writing and interpreting xTEDS. Participation in standard development is, however, restricted to U.S. organizations and a few international partners working under bi-lateral agreements.

From a technical point of view SPA has the potential to expand to a generic plug-and-play standard for robotics. Satellites include sensors, actuators, locomotion, navigation, on-board computing, energy management and communication just as advanced mobile robots. A broader ontology is needed and possibly an expanded set of communication specifications. The main hurdle is the restricted set of participants and the limited scope of the present effort.

The Joint Architecture for Unmanned Systems (JAUS) [17] was initiated by the U.S. Department of Defense and is presently developed by the Society of Automotive Engineers (SAE). JAUS takes a top-down service-oriented approach and covers presently two different levels of integration. The highest level specifies communication protocols between system components such as unmanned vehicles, payload and command and control systems. The second level handles interoperability between major subsystems of a mobile robot where each subsystem is understood to be a processing platform e.g. locomotion, sensor array and planning module. JAUS does not specify anything about architecture or implementation but is essentially a message-based protocol.

Many extensions including human-machine interfaces and manipulation services are drafted. JAUS has a large footprint in U.S. and NATO military robotics but is not applied extensively in other industrial or in academic contexts.

Conclusions

From this brief review of current standardization initiatives it is obvious that the robotics community is well aware of the need for standardization and that many relevant efforts are in progress. There is, however, no leading candidate for a global plug-and-play standard.

No industrial player is in the position of establishing a de facto standard as IBM once did for the PC market. Sony is too proprietary and Microsoft has at best a partial solution. Open source groups like ROS and Orca cater mostly to academics and fail to attract substantial groups of industrial users. OMG's Robotic Domain Taskforce and Space Plug-and-Play Avionics are both focused on component-level interoperability. The former is more abstract and address the global robotics industry including major players in Asia while the latter is technically more specific but limited to satellites and avionics. JAUS focuses on system-level integration and impacts mainly on U.S. and NATO military robotics.

Several of the emerging standards could be merged to form a more complete and powerful option. The JAUS high-level approach could be combined with the component and hardware oriented Space Plug-and-Play Avionics. Alternatively JAUS could be united with the OMG standards as suggested in [18].

Few computer scientists would claim that the IBM PC-architecture is optimal or even was the best option at the onset of the PC-revolution. Yet it transformed the industry and changed society with consequences that still reverberates. Similarly we should note that a plug-and-play robotics standard does not need to be technically optimal or respect all particular interests and opinions.

Many divergent forces resist standardization. Robot manufactures that are organized according to the vertical business model guard proprietary technologies. Academics are prone to prioritize technical optimization and will find faults in any proposed standard. Different industrial segments including space, defence, service, industrial and entertainment robotics have different requirements and traditions and there will always be arguments for specialization and adaption to the idiosyncrasies of the segment. Yet all parties would benefit enormously from a common standard. A strong integrative force such as government initiative might be needed for overcoming the diverging forces and catalysing a global standard spanning all relevant domains.

Establishing a robotics standard would not remove the need for hand-crafting military systems. For computers there is a segment for hardware and software that fulfils special military requirements of security, reliability, robustness and electromagnetic compatibility. Similar concerns will extend to robot components and systems.

Antennas and other electromagnetic emitters must be integrated with care and knowledge because of electromagnetic compatibility issues. Flying and sea-going robots must comply with aerodynamic and hydrodynamic constraints.

In spite of these and similar concerns it is obvious that a robotics plug-and-play standard would be instrumental for reducing cost and development time for advanced autonomous military systems.

Military robots are presently mainly operator-controlled and used for surveillance, de-mining and for combat in low-intensity conflicts. Driven by a rapid evolution of generic robot technology military robots will be increasingly autonomous, capable, affordable and available. This could well trigger an arms race and destabilize the present world order by enabling an unpredictable development and spread of innovative military technology. Aggressive second and third tier powers could, as demonstrated by plentiful historical examples, lead the march into the new era. The robotics transition in military affairs appears to be inevitable but perilous and could increase the risk of armed conflicts. Ethical concerns related to the new technology are also significant [19].

ORNITHOPTER

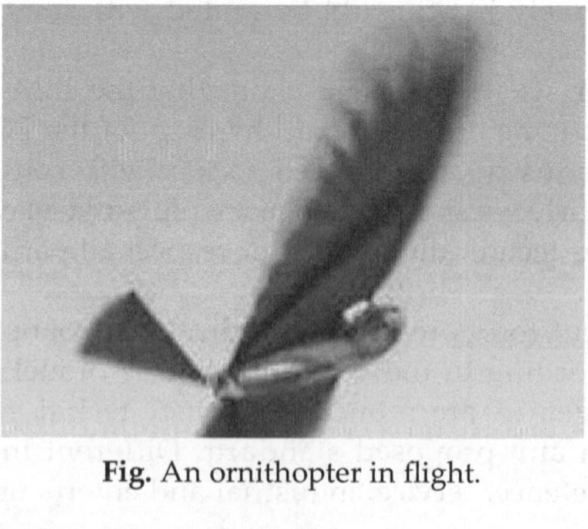

Fig. An ornithopter in flight.

Fig. Cybird radio-controlled ornithopter

An ornithopter is an aircraft that flies by flapping its wings. Designers seek to imitate the flapping-wing flight of birds, bats, and insects. Though machines may differ in form, they are usually built on the same scale as these flying creatures. Manned ornithopters have also been built, and some have been successful. The machines are of two general types: those with engines, and those powered by the muscles of the pilot.

EARLY HISTORY OF THE ORNITHOPTER

The Sanskrit epic Ramayana (4th Century BC) describes an ornithopter, the Pushpaka Vimana. The ancient Greek legend ofDaedalus (Greek demigod engineer) and Icarus (Daedalus's son) and The Chinese Book of Han (19 AD) both describe the use of feathers to make wings for a person but these are not actually aircraft. Some early manned flight attempts may have been intended to achieve flapping-wing flight though probably only a glide was actually achieved.

These include the flights of the 11th-century monk Eilmer of Malmesbury (recorded in the 12th century) and the 9th-century poet Abbas Ibn Firnas (recorded in the 17th century). Roger Bacon, writing in 1260, was also among the first to consider a technological means of flight. In 1485,Leonardo da Vinci began to study the flight of birds. He grasped that humans are too heavy, and not strong enough, to fly using wings simply attached to the arms. Therefore he sketched a device in which the aviator lies down on a plank and works two large, membranous wings using hand levers, foot pedals, and a system of pulleys.

The first ornithopters capable of flight were constructed in France. In 1858 Pierre Jullien's model flew an estimated forty feet.Gustave Trouvé's 1870 model flew a distance of 70 metres in a demonstration for the French Academy of Sciences. The wings were flapped by gunpowder charges activating a bourdon tube. Jobert in 1871 used a rubber band to power a small model bird.Alphonse Pénaud, Abel Hureau de Villeneuve, and Victor Tatin, also made rubber-powered ornithopters during the 1870s. Tatin's ornithopter (now in the US Air & Space Museum) was perhaps the first to use active torsion of the wings, and apparently it served as the basis for a commercial toy offered by Pichancourt c. 1889.

From 1884 on, Lawrence Hargrave built scores of ornithopters powered by rubber bands, springs, steam, or compressed air. He introduced the use of small flapping wings providing the thrust for a larger fixed wing. This eliminated the need for gear reduction, thereby simplifying the construction.

E.P. Frost made ornithopters starting in the 1870s, first models power by steam engines then in the 1900s an internal combustion one large enough for a person.

In the 1930s, Alexander Lippisch and the NSFK in Germany constructed and successfully flew a series of internal combustion powered ornithopters using a similar overall design, with aerodynamic improvements resulting from methodical study.

Erich von Holst also working in the 1930s, achieved great efficiency and realism in his work with ornithopters powered by rubber band. This includes perhaps the first success of an ornithopter with a bending wing, intended to more closely imitate the folding wing action of birds although it was not a true variable span wing like birds have.

MANNED FLIGHT

Manned ornithopters fall into two general categories: Those powered by the muscular effort of the pilot (human-powered ornithopters), and those powered by an engine. Around 1894, Otto Lilienthal, an aviation pioneer, became famous in Germany for his widely publicized and successful glider flights. Lilienthal also studied bird flight and conducted some related experiments. He constructed an ornithopter, although its complete development was prevented by his untimely death on the 9th of August 1896 in a glider accident.

In 1929, a man-powered ornithopter designed by Alexander Lippisch (designer of the Me163 Komet) flew a distance of 250 to 300 metres after tow launch. Since a tow launch was used, some have questioned whether the aircraft was capable of flying on its own. Lippisch asserted that the aircraft was actually flying, not making an extended glide. (Precise measurement of altitude and velocity over time would be necessary to resolve this question.) Most of the subsequent human-powered ornithopters likewise used a tow launch, and flights were brief simply because human muscle power diminishes rapidly over time.

In 1942, Adalbert Schmid made a much longer flight of a human-powered ornithopter at Munich-Laim. It travelled a distance of 900 metres, maintaining a height of 20 metres throughout most of the flight. Later this same aircraft was fitted with a 3 hp Sachs motorcycle engine. With the engine, it made flights up to 15 minutes in duration. Schmid later constructed a 10 hp ornithopter based on the Grunau-Baby IIa sailplane, which was flown in 1947. The second aircraft had flapping outer wing panels.

In 2005, Yves Rousseau was given the Paul Tissandier Diploma, awarded by the FAI for contributions to the field of aviation. Rousseau attempted his first human-muscle-powered flight with flapping wings in 1995. On 20 April 2006, at his 212th attempt, he succeeded in flying a distance of 64 metres, observed by officials of the Aero Club de France. Unfortunately, on his 213th flight attempt, a gust of wind led to a wing breaking up, causing the pilot to be gravely injured and rendered paraplegic.

A team at the University of Toronto Institute for Aerospace Studies, headed by Professor James DeLaurier, worked for several years on an engine-powered, piloted ornithopter. In July 2006, at the Bombardier Airfield at Downsview Park in Toronto, Professor DeLaurier's machine, the UTIAS Ornithopter No.1 made a jet-assisted takeoff and 14-second flight. According to DeLaurier, the jet was necessary for sustained flight, but the flapping wings did most of the work.

On August 2, 2010, Todd Reichert of the University of Toronto Institute for Aerospace Studies piloted a human-powered ornithopter named Snowbird. The 32 metres (105 ft 0 in) wingspan 42 kilograms (93 lb) aircraft was constructed from carbon fibre, balsa, and foam. The pilot sat in a small cockpit suspended below the wings and pumped a bar with his feet to operate a system of wires that flapped the wings up and down. Towed by a car until airborne, it then sustained flight for almost 20 seconds. It flew 145 meters with an average speed of 25.6 km/h (7.1 m/s) Similar tow-launched flights were made in the past, but improved data collection verified that the ornithopter was capable of self-powered flight once aloft.

APPLICATIONS FOR UNMANNED ORNITHOPTERS

Practical applications capitalize on the resemblance to birds or insects. The Colorado Division of Wildlife has used these machines to help save the endangered GunnisonSage Grouse. An artificial hawk under the control of an operator causes the grouse to remain on the ground so they can be captured for study.

Because ornithopters can be made to resemble birds or insects, they could be used for military applications, such as aerial reconnaissance without alerting the enemies that they are under surveillance. Several ornithopters have been flown with video cameras on board, some of which can hover and maneuver in small spaces. In 2011, AeroVironment, Inc. announced a remotely piloted ornithopter resembling a large hummingbird for possible spy missions.

AeroVironment, Inc., then led by Paul B. MacCready (Gossamer Albatross) developed in the mid-1980s, for the Smithsonian Institution, a half-scale radio controlled replica of the giant pterosaur, Quetzalcoatlus northropi. It was built to star in the IMAX movie On the Wing. The model had a wingspan of 5.5 metres (18 feet) and featured a complex,computerized autopilot control system, just as the full-size pterosaur relied on its neuromuscular system to make constant adjustments in flight.

Researchers hope to eliminate the motors and gears of current designs by more closely imitating animal flight muscles. Georgia Tech Research Institute's Robert C. Michelson is developing a Reciprocating Chemical Muscle for use in micro-scale flapping-wing aircraft. Michelson uses the term "entomopter" for this type of ornithopter.SRI International is developing polymer artificial muscles which may also be used for flapping-wing flight.

In 2002, Krister Wolff and Peter Nordin of Chalmers University of Technology in Sweden, built a flapping wing robot that learned flight techniques. The balsa wood design was driven by machine learning software technology known as a steady state linear evolutionary algorithm. Inspired by natural evolution, the software "evolves" in response to feedback on how well it performs a given task. Although confined to a laboratory apparatus, their ornithopter evolved behavior for maximum sustained lift force and horizontal movement. Since 2002, Prof. Theo van Holten has been working on an ornithopter which is constructed like a helicopter. The device is called the ornicopter and was made by constructing the main rotor so that it would have no reaction torque at all. In 2008, Schiphol Airport started using a real looking mechanical hawk designed by falconer Robert Musters. The radio controlled robot bird is used to scare away birds that could damage the engines of airplanes. In March 2011, scientists and engineers at the Festo Bionic Learning Network introduced a robotic SmartBird, based on the motion of a seagull. The SmartBird weighs only 450 grams and is controlled by a radio handset. On video, its flight appears remarkably realistic.

ORNITHOPTERS AS A HOBBY

Fig. The Dragonfly is a toy made by Wow-Wee.

Hobbyists can build and fly their own ornithopters. These range from light-weight models powered by rubber band, to larger models with radio control. The rubber-band-powered model can be fairly simple in design and construction. Hobbyists compete for the longest flight times with these models. An introductory model can be fairly simple in design and construction, but the advanced competition designs are extremely delicate and challenging to build. Roy White holds the United States national record for indoor rubber-powered, with his flight time of 21 minutes, 44 seconds.

Commercial free-flight rubber-band powered toy ornithopters have long been available. The first of these was sold under the nameTim Bird in Paris in 1879. Later models were also sold as Tim Bird (made by G de Ruymbeke, France, since 1969).

Commercial radio controlled designs stem from Percival Spencer's engine-powered Seagulls, developed circa 1958, and Sean Kinkade's work in the late 1990s to present day. The wings are usually driven by an electric motor. Many hobbyists enjoy experimenting with their own new wing designs and mechanisms. The opportunity to interact with real birds in their own domain also adds great enjoyment to this hobby. Birds are often curious and will follow or investigate the model while it is flying. In a few cases, RC birds have been attacked by birds of prey, crows, and even cats. More recent cheaper models such as the Dragonfly from WowWee have extended the market from dedicated hobbyists to the general toy market.

Some helpful resources for hobbyists include The Ornithopter Design Manual, book written by Nathan Chronister, and The Ornithopter Zone web site, which includes a large amount of information about building and flying these models. To see video examples of a remote control Ornithopter visit the Birds You Fly website.

Ornithopters are also of interest as the subject of one of the events in the nationwide Science Olympiad event list. The event ("Flying Bird") entails building a self-propelled ornithopter to exacting specifications, with points awarded for high flight time and low weight. Bonus points are also awarded if the ornithopter happens to look like a real bird.

AERODYNAMICS

As demonstrated by birds, flapping wings offer potential advantages in maneuverability and energy savings compared with fixed-wing aircraft, as well as potentially vertical take-off and landing. It has been suggested that these advantages are greatest at small sizes and low flying speeds.

Unlike airplanes and helicopters, the driving airfoils of the ornithopter have a flapping or oscillating motion, instead of rotary. As with helicopters, the wings usually have a combined function of providing both lift and thrust. Theoretically, the flapping wing can be set to zero angle of attack on the upstroke, so it passes easily through the air. Since typically the flapping airfoils produce both lift and thrust, drag-inducing structures are minimized. These two advantages potentially allow a high degree of efficiency.

WING DESIGN

An effective ornithopter must have wings capable of generating both thrust, the force that propels the craft forward, and lift, the force, perpendicular to the direction of flight, that keeps the craft airborne. These forces must be strong enough to counter the effects of drag and the weight of the craft. Leonardo's ornithopter designs were inspired by his study of birds, and conceived the use of flapping motion to generate thrust and provide the forward motion necessary for aerodynamic lift. However, using materials available at that time, the craft would be too heavy and require too much energy to produce sufficient lift or thrust for flight.

Alphonse Pénaud introduced the idea of a powered ornithopter in 1874. His design had limited power and was uncontrollable, causing it to be transformed into a toy for children. More recent vehicles, such as the human-powered ornithopters of Lippisch (1929) and Emil Hartman (1959), were capable powered gliders, but required a towing vehicle in order to take off, and they may not have been capable of generating sufficient lift for sustained flight. Hartman's ornithopter lacked the theoretical background of others based on the study of winged flight, but exemplified the idea of an ornithopter as a birdlike machine, rather than a machine that directly copies birds' method of flight.

The 1960s saw powered, unmanned ornithopters of various sizes capable of achieving and sustaining flight, providing valuable real-world examples of mechanical, winged flight. In 1991, Harris and DeLaurier flew the first successful engine-powered remotely piloted ornithopter in Toronto, Canada. In 1999, a piloted ornithopter based on this design flew, capable of taking off from level pavement and sustained flight.

An ornithopter's flapping wings and their motion through the air are designed to maximize the amount of lift generated within limits of weight, material strength, and mechanical complexity. A flexible wing material can increase efficiency while keeping the driving mechanism simple. In wing designs with the spar sufficiently forward of the airfoil that the aerodynamic center is aft of the elastic axis of the wing, aeroelastic deformation causes the wing to move in a manner close to its ideal efficiency (in which pitching angles lag plunging displacements by approximately 90 degrees.) Flapping wings increase drag and are not as efficient as propeller-powered aircraft. Some designs achieve increased efficiency by applying more power on the down stroke than on the upstroke.

In order to achieve the desired flexibility and minimum weight, engineers and researchers have experimented with wings that require carbon fiber, plywood, fabric, and ribs, with a stiff, strong trailing edge. Any mass located to the aft of the empennage reduces the wing's performance, so lightweight materials and empty space are used where possible. In order to minimize drag and maintain the desired shape, choice of a material for the wing surface is also important. In DeLaurier's experiments, a smooth aerodynamic surface with a double-surface airfoil is more efficient at producing lift than a single-surface airfoil.

Other ornithopters do not necessarily act like birds or bats in flight. Typically birds and bats have thin and cambered wings to produce lift and thrust. Ornithopters with thinner wings have a limited angle of attack but provide optimum minimum-drag performance for a single lift coefficient.

Although hummingbirds fly with fully extended wings, such flight is not feasible for an ornithopter. If an ornithopter wing were to fully extend and twist and flap in small movements it would cause a stall, and if it were to twist and flap in very large motions, it would act like a windmill causing an inefficient flying situation.

A team of engineers and researchers called "Fullwing" has created an ornithopter that has an average lift of over 8 pounds, an average thrust of 0.88 pounds, and has a propulsive efficiency of 54%. The wings were tested in a low speed wind tunnel measuring the aerodynamic performance, discovering that the higher the frequency of the wing beat, the higher the average thrust of the ornithopter.

UNMANNED AERIAL VEHICLE

Fig. A group photo of aerial demonstrators at the 2005 Naval Unmanned Aerial Vehicle Air Demo.

An unmanned aerial vehicle (UAV), also known as a drone, is an aircraft without a human pilot on board. Its flight is controlled either autonomously by computers in the vehicle or under the remote control of a pilot on the ground or in another vehicle. The typical launch and recovery method or "Pop and Stop" of an unmanned aircraft is by the function of an automatic system or an external operator on the ground. There are a wide variety of UAV shapes, sizes, configurations, and characteristics. Historically, UAVs were simple remotely piloted aircraft, but autonomous control is increasingly being employed.

They are usually deployed for military and special operation applications, but also used in a small but growing number of civil applications, such as policing and firefighting, and nonmilitary security work, such as surveillance of pipelines. UAVs are often preferred for missions that are too "dull, dirty or dangerous" for manned aircraft.

HISTORY

Fig. Ryan Firebee was a series of target drones/unmanned aerial vehicles.

The earliest attempt at a powered unmanned aerial vehicle was A. M. Low's "Aerial Target" of 1916. Nikola Tesla described a fleet of unmanned aerial combat vehicles in 1915. A number of remote-controlled airplane advances followed during and afterWorld War I, including the Hewitt-Sperry Automatic Airplane. The first scale RPV (Remote Piloted Vehicle) was developed by the film star and model airplane enthusiast Reginald Denny in 1935.

More were made in the technology rush during World War II; these were used both to train antiaircraft gunners and to fly attack missions. Nazi Germany also produced and used various UAV aircraft during the course of WWII. Jet engines were applied after World War II in such types as the Teledyne Ryan Firebee I of 1951, while companies like Beechcraft also got in the game with their Model 1001 for the United States Navy in 1955.Nevertheless, they were little more than remote-controlled airplanes until the Vietnam Era.

The birth of U.S. UAVs (called RPVs at the time) began in 1959 when United States Air Force (USAF) officers, concerned about losing pilots over hostile territory, began planning for the use of unmanned flights. This plan became intensified when Francis Gary Powers and his "secret" U-2 were shot down over the Soviet Union in 1960. Within days, the highly classified UAV program was launched under the code name of "Red Wagon". The August 2 and August 4, 1964, clash in the Tonkin Gulf between naval units of the U.S. and North Vietnamese Navy initiated America's highly classified UAVs into their first combat missions of the Vietnam War. When the "Red Chinese" showed photographs of downed U.S. UAVs viaWide World Photos, the official U.S. response was "no comment."

There are two prominent UAV programs within the United States: that of the military and that of the Central Intelligence Agency (CIA). The military's UAV program is overt, meaning it is recognized by the public and therefore only operates where US troops are stationed. The CIA's program is covert. Missions performed by the CIA's UAV program do not always occur where US troops are stationed.

Fig. The Israeli Tadiran Mastiff, which first flew in 1973, is seen as the first modern battlefield UAV, due to its data-link system, endurance-loitering, and live video streaming.

The CIA's UAV program was commissioned as a result of the September 11 terrorist attacks and the increasing emphasis on operations for intelligence gathering in 2004. This clandestine program is primarily being used in Afghanistan, Pakistan, Yemen, and Somalia. UAVs collect intelligence in these countries by loitering around their target. The CIA's first UAV program is called the Eagle Program. It was led by Duane Clarridge, the director of the Counterterrorism Center. This program constructed the CIA's first using "off the shelf technology," which included items such as garage door openers and model airplanes.

Only on February 26, 1973, during testimony before the United States House Committee on Appropriations, the U.S. military officially confirmed that they had been utilizing UAVs in Southeast Asia (Vietnam). Over 5,000 U.S. airmen had been killed and over 1,000 more were either missing in action (MIA) or captured (prisoners of war/POW). The USAF 100th Strategic Reconnaissance Wing had flown approximately 3,435 UAV missions during the war at a cost of about 554 UAVs lost to all causes. In the words of USAF General George S. Brown, Commander, Air Force Systems Command, in 1972, "The only reason we need (UAVs) is that we don't want to needlessly expend the man in the cockpit." Later that same year, General John C. Meyer, Commander in Chief, Strategic Air Command, stated, "we let the drone do the high-risk flying... the loss rate is high, but we are willing to risk more of them... they save lives!"

During the 1973 Yom Kippur War, Soviet-supplied surface-to-air missile batteries in Egypt and Syria caused heavy damage to Israeli fighter jets. As a result, Israel developed the first modern UAV. Israel pioneered the use of UAVs for real-time surveillance, electronic warfare, and decoys. The images and radar decoying provided by these UAVs helped Israel to completely neutralize the Syrian air defenses at the start of the 1982 Lebanon War, resulting in no pilots downed. The first time UAVs were used as proof-of-concept of super-agility post-stall controlled flight in combat flight simulations was with tailless, stealth technology-based, three-dimensional thrust vectoring flight control, jet steering UAVs in Israel in 1987.

With the maturing and miniaturization of applicable technologies as seen in the 1980s and 1990s, interest in UAVs grew within the higher echelons of the U.S. military. In the 1990s, the U.S. Department of Defense gave a contract to U.S. corporation AAI Corporation of Maryland along with Israeli company Mazlat.

The U.S. Navy bought the AAI Pioneer UAV that was jointly developed by American AAI Corporation and Israeli Mazlat, and this type of UAV is still in use. Many of these Pioneer and newly developed U.S. UAVs were used in the 1991 Gulf War. UAVs were seen to offer the possibility of cheaper, more capable fighting machines that could be used without risk to aircrews. Initial generations were primarily surveillance aircraft, but some were armed, such as the General Atomics MQ-1 Predator, which utilized AGM-114 Hellfire air-to-ground missiles. An armed UAV is known as an unmanned combat air vehicle (UCAV).

As a tool for search and rescue, UAVs can help find humans lost in the wilderness, trapped in collapsed buildings, or adrift at sea.

In February 2013, it was reported that UAVs were used by at least 50 countries, several of which made their own: for example, Iran, Israel and China.

The first UAV created was the Pioneer, which helps to identify artillery and boats (Carafano & Gudgel, 2007). Since its performance was so exemplary, they began to be used more and more, with new models constantly being introduced. As of 2008, the United States Air Force employed 5,331 UAVs, which is twice its number of manned planes.Out of these, the Predators are the most commendable.

Unlike other UAVs, the Predator was armed with Hellfire missiles so that it can terminate the target that it locates (Carafano & Gudgel, 2007). This was done after Predators sighted Osama Bin Laden multiple times but could not do anything about it other than send back images. In addition, the Predator is capable of orchestrating attacks by pointing lasers at the targets (Singer,

2009b). This is important, as it puts a robot in a position to set off an attack. Their overall success is apparent because from June 2005 to June 2006 alone, Predators carried out 2,073 missions and participated in 242 separate raids (Singer, 2009a).

In contrast to the Predator, which is remotely piloted via satellites by pilots located 7,500 miles away, the Global Hawk operates virtually autonomously. The user merely hits the button for 'take off' and for 'land', while the UAV gets directions via GPS and reports back with a live feed. Global Hawks have the capability to fly from San Francisco and map out the entire state of Maine before having to return. In addition, some UAVs have become so small that they can be launched from one's hand and maneuvered through the street.

These UAVs, known as Ravens, are especially useful in urban areas, such as Iraq, in order to discover insurgents and potential ambushes the next block up (Carafano & Gudgel, 2007). UAVs are especially useful because they can fly for days at a time. According to Carafano & Gudgel, insurgents are loathe to stay in the open for more than a few minutes at a time for fear of UAVs locating them (2007).

FAA DESIGNATION

In the United States, the United States Navy and, shortly after, the Federal Aviation Administration have adopted the name unmanned aircraft (UA) to describe aircraft systems without a flight crew on board. More common names include UAV, drone, remotely piloted vehicle (RPV), remotely piloted aircraft (RPA), and remotely operated aircraft (ROA). These "limited-size" (as defined by the FAI) unmanned aircraft flown in the USA's National Airspace System, flown solely for recreation and sport purposes, such as models, are generally flown under the voluntary safety standards of the Academy of Model Aeronautics, the United States' national aeromodeling organization.

To operate a UA for non-recreational purposes in the United States, users must obtain a Certificate of Authorization (COA) to operate in national airspace.

At the moment, COAs require a public entity as a sponsor. For example, when BP needed to observe oil spills, they operated the Aeryon Scout UAVs under a COA granted to the University of Alaska Fairbanks. COAs have been granted for both land and shipborne operations.

As of August 2013, commercial unmanned aerial system (UAS) licenses were granted on a case-by-case basis, subject to approval by the Federal Aviation Administration (FAA). A Congressional mandate to integrate UASs into U.S. airspace protocols is forecast to grant FAA licenses more broadly as early as 2015, the agency expecting that five years after it unveils a regulatory framework for UASs weighing 55 pounds or less, there will be 7,500 such devices in the air.

The term unmanned aircraft system (UAS) emphasizes the importance of other elements beyond an aircraft itself. A typical UAS consists of the following:

- Unmanned aircraft (UA);
- Control system, such as Ground Control Station (GCS);
- Control link, a specialized datalink; and
- Other related support equipment.

For example, the RQ-7 Shadow UAS consists of four UAs, two GCSs, one portable GCS, one Launcher, two Ground Data Terminals (GDTs), one portable GDT, and one Remote Video Terminal. Certain military units are also fielded with a maintenance support vehicle.

Because of this systemic approach, unmanned aircraft systems have not been included in the United States Munitions List Category VIII – Aircraft and Associated Equipment. Vice versa, the "Unmanned Aerial Vehicle Systems" are clearly mentioned at paragraph 121-16 Missile Technology Control Regime Annex of the United States Munitions List. More precisely, the Missile Technology Control Regime Annex levels rocket and unmanned aerial vehicle systems together.

The term UAS was since adopted by the United States Department of Defense (DOD) and the British Civil Aviation Authority (CAA).

The term used previously for unmanned aircraft system was unmanned-aircraft vehicle system (UAVS).

CLASSIFICATION

Fig. Although most UAVs are fixed-wing aircraft, rotorcraft designs (i.e., RUAVs) such as this MQ-8B Fire Scout are also used.

UAVs typically fall into one of six functional categories (although multi-role airframe platforms are becoming more prevalent):

- Target and decoy – providing ground and aerial gunnery a target that simulates an enemy aircraft or missile

- Reconnaissance – providing battlefield intelligence
- Combat – providing attack capability for high-risk missions (see Unmanned combat air vehicle)
- Logistics – UAVs specifically designed for cargo and logistics operation
- Research and development – used to further develop UAV technologies to be integrated into field deployed UAV aircraft
- Civil and Commercial UAVs – UAVs specifically designed for civil and commercial applications

Fig. Schiebel S-100 fitted with a Lightweight Multirole Missile

They can also be categorised in terms of range/altitude and the following has been advanced as relevant at such industry events as ParcAberporth Unmanned Systems forum:

- Hand-held 2,000 ft (600 m) altitude, about 2 km range
- Close 5,000 ft (1,500 m) altitude, up to 10 km range
- NATO type 10,000 ft (3,000 m) altitude, up to 50 km range
- Tactical 18,000 ft (5,500 m) altitude, about 160 km range
- MALE (medium altitude, long endurance) up to 30,000 ft (9,000 m) and range over 200 km
- HALE (high altitude, long endurance) over 30,000 ft (9,100 m) and indefinite range
- HYPERSONIC high-speed, supersonic (Mach 1–5) or hypersonic (Mach 5+) 50,000 ft (15,200 m) or suborbital altitude, range over 200 km
- ORBITAL low earth orbit (Mach 25+)
- CIS Lunar Earth-Moon transfer
- CACGS Computer Assisted Carrier Guidance System for UAVs

The United States military employs a tier system for categorizing its UAVs.

Classifications by the United States military

The modern concept of U.S. military UAVs is to have the various aircraft systems work together in support of personnel on the ground. The integration scheme is described in terms of a "Tier" system and is used by military planners to designate the various individual aircraft elements in an overall usage plan for integrated operations. The Tiers do not refer to specific models of aircraft but rather roles for which various models and their manufacturers competed. The U.S. Air Force and the U.S. Marine Corps each has its own tier system, and the two systems are themselves not integrated.

U.S. Air Force tiers

Fig. An MQ-9 Reaper, a hunter-killersurveillance UAV.

- Tier N/A: Small/Micro UAV. Role filled by BATMAV (Wasp Block III).
- Tier I: Low altitude, long endurance. Role filled by the Gnat 750.
- Tier II: Medium altitude, long endurance (MALE). Role currently filled by the MQ-1 Predator and MQ-9 Reaper.
- Tier II+: High altitude, long endurance conventional UAV (or HALE UAV). Altitude: 60,000 to 65,000 feet (19,800 m), less than 300 knots (560 km/h) airspeed, 3,000-nautical-mile (6,000 km) radius, 24 hour time-on-station capability. Complementary to the Tier III- aircraft. Role currently filled by the RQ-4 Global Hawk.
- Tier III-: High altitude, long endurance low-observable UAV. Same parameters as, and complementary to, the Tier II+ aircraft. The RQ-3 DarkStar was originally intended to fulfill this role before it was "terminated". Role now filled by RQ-170 Sentinel.

U.S. Marine Corps tiers

- Tier N/A: Micro UAV. Wasp III fills this role, driven largely by the desire for commonality with the USAF BATMAV.
- Tier I: Role currently filled by the Dragon Eye but all ongoing and future procurement for the Dragon Eye program is going now to the RQ-11B Raven B.
- Tier II: Role currently filled by the ScanEagle.

- Tier III: For two decades, the role of medium range tactical UAV was filled by the Pioneer UAV. In July 2007, the Marine Corps announced its intention to retire the aging Pioneer fleet and transition to the RQ-7 Shadow Tactical Unmanned Aircraft System by AAI Corporation. The first Marine Shadow systems have already been delivered, and training for their respective Marine Corps units is underway.

U.S. Army tiers

- Tier I: Small UAV. Role filled by the RQ-11B Raven.
- Tier II: Short Range Tactical UAV. Role filled by the RQ-7B Shadow 200.
- Tier III: Medium Range Tactical UAV. Role currently filled by the MQ-5A/B Hunter and IGNAT/IGNAT-ER, but transitioning to the Extended Range Multi-Purpose (ERMP)MQ-1C Gray Eagle.

Future Combat Systems (FCS) (U.S. Army) classes

- Class I: For small units. Role to be filled by all new UAV with some similarity to micro air vehicle.
- Class II: For companies (cancelled).
- Class III: For battalions (cancelled).
- Class IV: For brigades. Role to be filled by the RQ-8A/B/ MQ-8B Fire Scout.

Conversions or variants of existing manned aircraft

- A-10PCAS, a Fairchild Republic A-10 Thunderbolt II variant (in development).

Unmanned aircraft system

An unmanned aircraft system (UAS) includes ground stations and other elements besides the actual aircraft. The term was first officially used by the FAA in early 2005 and subsequently adopted by DoD that same year in their Unmanned Aircraft System Roadmap 2005–2030. Many people have mistakenly used the term Unmanned AerialSystem or Unmanned Air Vehicle System, as these designations were in provisional use at one time or another.

The inclusion of the term aircraft emphasizes that regardless of the location of the pilot and flightcrew, the operations must comply with the same regulations and procedures as do those aircraft with the pilot and flightcrew on board.

The official acronym UAS is also used by the International Civil Aviation Organization (ICAO) and other government aviation regulatory organizations.

Fig. Predator launching a Hellfire missile

The military role of unmanned aircraft systems is growing at unprecedented rates. In 2005, tactical- and theater-level unmanned aircraft alone had flown over 100,000 flight hours in support of Operation Enduring Freedom and Operation Iraqi Freedom, in which they are organized under Task Force Liberty in Afghanistan and Task Force ODIN in Iraq. Rapid advances in technology are enabling more and more capability to be placed on smaller airframes, which is spurring a large increase in the number of Small Unmanned Aircraft Systems (SUAS) being deployed on the battlefield.

The use of SUAS in combat is so new that no formal DoD wide reporting procedures have been established to track SUAS flight hours. As the capabilities grow for all types of UAS, nations continue to subsidize their research and development, leading to further advances and enabling them to perform a multitude of missions. UAS no longer only perform intelligence, surveillance, and reconnaissance missions, although this still remains their predominant type.

Their roles have expanded to areas including electronic attack, strike missions, suppression and/or destruction of enemy air defense, network node or communications relay, combat search and rescue, and derivations of these themes. These UAS range in cost from a few thousand dollars to tens of millions of dollars, with aircraft ranging from less than one pound to over 40,000 pounds.

When the Obama administration announced in December 2009 the deployment of 30,000 new troops in Afghanistan, there was already an increase of attacks by unmanned Predator UAVs against Taliban and al-Qaeda militants in Afghanistan and Pakistan's tribal areas, of which one probably killed a key member of al-Qaeda. However, neitherOsama bin Laden nor Ayman al-Zawahiri was the likely target, according to reports.

According to a report of the New America Foundation, armed UAV strikes had dramatically increased under President Obama – even before his deployment decision. There were 43 such attacks between January and October 2009. The report draws on what it deems to be "credible" local and national media stories about the attacks. This can be compared to a total of 34 in all of 2008, which was President Bush's last full year in office. Between 2006 and

2009, UAV-launched missiles allegedly had killed between 750 and 1,000 people in Pakistan, according to the report. Of these, about 20 people were said to be leaders of al-Qaeda, Taliban, and associated groups. Overall, 66% to 68% of the people killed were militants, and 31% to 33% were civilians. U.S. officials disputed the percentage for civilians. The U.S. Air Force has recently begun referring at least to larger UAS like Predator, Reaper, and Global Hawk as Remotely Piloted Aircraft (RPA) to highlight the fact that these systems are always controlled by a human operator at some location.

However, artificial intelligence is advancing to the point where the aircraft are easily capable of taking off, landing, and flying themselves. Then they simply have to be instructed as to their mission. The military distinguishes between "man in the loop"(piloted) and "man on the loop" (supervised) systems, with "fully autonomous" (issued orders) growing organically from the second into a third category. A.I. systems have been capable of making decisions and planning sequences of actions for decades; as of 2013, few fully autonomous systems have been constructed, but this is more a matter of convenience and technical implementation than of any fundamental barrier.

To distinguish UAVs from missiles, a UAV is defined as a "powered, aerial vehicle that does not carry a human operator, uses aerodynamic forces to provide vehicle lift, can fly autonomously or be piloted remotely, can be expendable or recoverable, and can carry a lethal or nonlethal payload". Therefore, cruise missiles are not considered UAVs because, like many other guided missiles, the vehicle itself is a weapon that is not reused, even though it is also unmanned and in some cases remotely guided.

Uses

Fig. A thermal imaging gimbal pod camera mounted on the side of a Huey UH-1, similar to those used on certain UAVs.

Fig. Aeryon Scout in flight.

Fig. An InView UAV used in various scientific and commercial applications.

Fig. The RQ-7 Shadow is capable of delivering a 20 lb (9.1 kg) "Quick-MEDS" canister to front-line troops.

Fig. Fulmar UAV, developed by Aerovision for civilian applications.

Fig. IAI Heron, an Unmanned Aerial Vehicle developed by the Malat (UAV) division ofIsrael Aerospace Industries.

Fig. A UAV detecting an underground facility

Fig. A Mexican S4 Ehécatl for a surveillance mission in a take-off

Beyond the military applications of UAVs with which "drones" became most associated, numerous civil aviation uses have been developed, including aerial surveying of crops, acrobatic aerial footage in filmmaking, search and rescue operations, inspecting power lines and pipelines, and counting wildlife, with manufacturers rebranding the technology as "unmanned aerial systems" (UASs) in preference over "drones."

Remote sensing

UAV remote sensing functions include electromagnetic spectrum sensors, gamma ray sensors, biological sensors, and chemical sensors. A UAV's electromagnetic sensors typically include visual spectrum, infrared, or near infrared cameras as well as radar systems. Other electromagnetic wave detectors such as microwave and ultraviolet spectrum sensors may also be used but are uncommon. Biological sensors are sensors capable of detecting the airborne presence of various microorganisms and other biological factors. Chemical sensors use laser spectroscopy to analyze the concentrations of each element in the air.

Commercial aerial surveillance

Aerial surveillance of large areas is made possible with low cost UAV systems. Surveillance applications include livestock monitoring, wildfire mapping, pipeline security, home security, road patrol, and anti-piracy. The trend for the use of UAV technology in commercial aerial surveillance is expanding rapidly with increased development of automated object detection approaches.

Domestic policing

UAVs are increasingly used for domestic police work in Canada and the

United States: a dozen US police forces had applied for UAV permits by March 2013. Texas politician and commentator Jim Hightower has warned about potential privacy abuses from aerial surveillance. In February 2013, Seattle Mayor Michael McGinn responded to protests by scrapping the Seattle Police Department's plan to deploy UAVs.

Oil, gas and mineral exploration and production

UAVs can be used to perform geophysical surveys, in particular geomagnetic surveys where the processed measurements of the Earth's differential magnetic field strength are used to calculate the nature of the underlying magnetic rock structure. A knowledge of the underlying rock structure helps trained geophysicists to predict the location of mineral deposits.

The production side of oil and gas exploration and production entails the monitoring of the integrity of oil and gas pipelines and related installations. For above-ground pipelines, this monitoring activity could be performed using digital cameras mounted on one or more UAVs. The InView Unmanned Aircraft System is an example of a UAV developed for use in oil, gas, and mineral exploration and production activities.

Transport

UAVs can transport goods using various means based on the configuration of the UAV itself. Most payloads are stored in an internal payload bay somewhere in the airframe. For many helicopter configurations, external payloads can be tethered to the bottom of the airframe. With fixed-wing UAVs, payloads can also be attached to the airframe, but aerodynamics of the aircraft with the payload must be assessed. For such situations, payloads are often enclosed in aerodynamic pods for transport.

Scientific research

Unmanned aircraft are especially useful in penetrating areas that may be too dangerous for manned aircraft. The National Oceanic and Atmospheric Administration (NOAA) began utilizing the Aerosonde unmanned aircraft system in 2006 as a hurricane hunter. AAI Corporation subsidiary Aerosonde Pty Ltd. of Victoria, Australia, designs and manufactures the 35-pound system, which can fly into a hurricane and communicate near-real-time data directly to the National Hurricane Center in Florida.

Beyond the standard barometric pressure and temperature data typically culled from manned hurricane hunters, the Aerosonde system provides measurements far closer to the water's surface than previously captured. Further applications for unmanned aircraft can be explored once solutions have been developed for their accommodation within national airspace, an issue currently under discussion by the Federal Aviation Administration.

UAVSI, the UK manufacturer, also produces a variant of their Vigilant light UAS (20 kg) designed specifically for scientific research in severe climates, such as the Antarctic.

There have also been experiments with using UAVs as a construction and artwork tool at locations such as the ETH Zurich.

Armed attacks

MQ-1 Predator UAVs armed with Hellfire missiles are increasingly used by the U.S. as platforms for hitting ground targets. Armed Predators were first used in late 2001 from bases in Pakistan and Uzbekistan, mostly aimed at assassinating high profile individuals (terrorist leaders, etc.) inside Afghanistan. Since then, there have been many reported cases of such attacks taking place in Afghanistan, Pakistan, Yemen, and Somalia. The advantage of using an unmanned vehicle rather than a manned aircraft in such cases is to avoid a diplomatic embarrassment should the aircraft be shot down and the pilots captured, since the bombings take place in countries deemed friendly and without the official permission of those countries.

A Predator based in a neighboring Arab country was used to kill suspected al-Qaeda terrorists in Yemen on November 3, 2002. This marked the first use of an armed Predator as an attack aircraft outside of a theater of war such as Afghanistan.

The U.S. has claimed that the Predator strikes killed at least nine senior al-Qaeda leaders and dozens of lower-ranking operatives, depleting its operational tier in what U.S. officials described as the most serious disruption of al-Qaeda since 2001. It was claimed that the Predator strikes took such a toll on al-Qaeda that militants began turning violently on one another out of confusion and distrust. A senior U.S. counter-terrorism official said: "They have started hunting down people who they think are responsible" for security breaches. "People are showing up dead, or disappearing."

By October 2009, the CIA claimed to have killed more than half of the 20 most wanted al-Qaeda terrorist suspects in targeted killings using UAVs. By May 2010, counter-terrorism officials said that UAV strikes in the Pakistani tribal areas had killed more than 500 militants since 2008 and no more than 30 (5%) nearby civilians—mainly family members who lived and traveled with the targets. UAVs linger overhead after a strike, in some cases for hours, to enable the CIA to count the bodies and attempt to determine which, if any, are civilians. A Pakistani intelligence officer gave a higher estimate of civilian casualties, saying 20% of total deaths were civilians or non-combatants.

In February 2013, U.S. Senator Lindsey Graham stated that 4,756 people have been killed by U.S. UAVs.

CIA officials became concerned in 2008 that targets in Pakistan were being tipped off to pending U.S. UAV strikes by Pakistani intelligence when the U.S. requested Pakistani permission prior to launching UAV-based attacks.

The Bush administration therefore decided in August 2008 to abandon the practice of obtaining Pakistani government permission before launching missiles from UAVs, and in the next six months the CIA carried out at least 38 Predator strikes in northwest Pakistan, compared with 10 in 2006 and 2007 combined.

In 2012, the USAF trained more UAV pilots than ordinary jet fighter pilots for the first time.

One issue with using armed drones to attack human targets is the size of the bombs being used and the relative lack of discrimination of the 100 lb (45 kg) Hellfire, which was designed to eliminate tanks and attack bunkers. Smaller weapons such as the Raytheon Griffin and Small Tactical Munition are being developed as a less indiscriminate alternative, and development is underway on the still smaller US Navy-developed Spike missile. The payload-limited Predator A can also be armed with six Griffin missiles, as opposed to only two of the much-heavier Hellfires.

Civilian casualties

Questions have been raised about the accuracy of UAV-based missile strikes. In March 2009, The Guardian reported allegations that Israeli UAVs armed with missiles killed 48 Palestinian civilians in the Gaza Strip, including two small children in a field and a group of women and girls in an otherwise empty street.

In June, Human Rights Watch investigated six UAV attacks that were reported to have resulted in civilian casualties and alleged that Israeli forces either failed to take all feasible precautions to verify that the targets were combatants or failed to distinguish between combatants and civilians. In July 2009, Brookings Institution released a report stating that in the United States-led drone attacks in Pakistan, ten civilians died for every militant killed. S. Azmat Hassan, a former ambassador of Pakistan, said in July 2009 that American UAV attacks were turning Pakistani opinion against the United States and that 35 or 40 such attacks only killed 8 or 9 top al-Qaeda operatives.

Although it may never be known how many civilians have died as a result of U.S. UAV strikes in Pakistan, there are estimates of hundreds or thousands of innocent bystanders who have perished in such attacks. Pakistani authorities released statistics indicating that between January 1, 2009 and December 31, 2009, U.S. RQ-1 Predator and RQ-9 Reaper UAV strikes have killed over 700 innocent civilians. The website PakistanBodyCount.Org (by Zeeshan-ul-hassan Usmani, a Fulbright Scholar at theFlorida Institute of Technology) shows 1,065 civilian deaths between June 2004 and January 30, 2010 and tallies 103 UAV strikes carried out by the United States.

With the increase of UAV strikes, January 2010 proved to be a deadly month in Pakistan with 123 innocent civilians killed, according to a story in

The International News. In addition, it has been reported that 160 children have died from UAV-launched attacks in Pakistan. Further, over 1,000 civilians have been injured. This evidence runs counter to the Obama administration's claim that "nearly for the past year there hasn't been a single collateral death" due to UAV-based attacks.

According to the February 24, 2010 policy analysis "The Year of the Drone" released by the New America Foundation, the civilian fatality rate since 2004 is approximately 32%. The study reports that 114 reported UAV-based missile strikes in northwest Pakistan from 2004 to present killed between 830 and 1,210 individuals, around 550 to 850 of whom were militants.

After more than 30 UAV-based strikes hit civilian homes in Afghanistan in 2012, President Hamid Karzai demanded that such attacks end, but the practice continues in areas of Pakistan, Yemen, and Somalia that are not in war zones. Former U.S. President Jimmy Carter has criticized such use of UAVs: "We don't know how many hundreds of innocent civilians have been killed in these attacks...This would have been unthinkable in previous times."

Aerial target practice in training of human pilots

Since 1997, the U.S. military has used more than 80 F-4 Phantoms converted into robotic planes for use as aerial targets for combat training of human pilots. The F-4s were supplemented in September 2013 with F-16s as more realistically maneuverable targets.

Search and rescue

UAVs will likely play an increased role in search and rescue in the United States. This was demonstrated by the use of UAVs during the 2008 hurricanes that struck Louisiana and Texas. Micro UAVs, such as the Aeryon Scout, have been used to perform Search and Rescue activities on a smaller scale, such as the search for missing persons.For example, Predators, operating between 18,000–29,000 feet above sea level, performed search and rescue and damage assessment. Payloads carried were an optical sensor, which is a daytime and infrared camera in particular, and a synthetic aperture radar (SAR). The Predator's SAR is a sophisticated all-weather sensor capable of providing photographic-like images through clouds, rain or fog, and in daytime or nighttime conditions, all in real-time. A concept of coherent change detection in SAR images allows for exceptional search and rescue ability: photos taken before and after the storm hits are compared, and a computer highlights areas of damage.

Conservation

In June 2012, WWF announced it will begin using UAVs in Nepal to aid conservation efforts following a successful trial of two aircraft in Chitwan National Park, with ambitions to expand to other countries, such as Tanzania and Malaysia. The global wildlife organization plans to train ten personnel to use the UAVs, with operational use beginning in the fall. In August 2012, UAVs were used by members of the Sea Shepherd Conservation Society in Namibia to document the annual seal cull. In March 2013, the Times published a controversial story that UAV conservation nonprofit ShadowView, founded by former members of Sea Shepherd Conservation Society, had been working for several months with anti-hunting charity The League Against Cruel Sports to expose illegal fox hunting in the UK; hunt supporters have argued that using UAVs to film hunting is an invasion of privacy.

Maritime patrol

Japan is studying how to deal with the UAVs the PRC is starting to use to enforce their claims on unmanned islands.

Forest fire detection

Another application of UAVs is the prevention and early detection of forest fires. The chief exponent of this type is the FT-ALTEA, developed by Flightech Systems.

The possibility of constant flight, both day and night, makes the methods used until now (helicopters, watchtowers, etc.) become obsolete. Its payload consists of numerous cameras (HD, thermal, hyperspectral, etc.) and multiple sensors that provide real-time emergency services, including information about the location of the outbreak of fire as well as many factors (wind speed, temperature, humidity, etc.) that are helpful for fire crews to conduct fire suppression.

Archaeology

In Peru archaeologists use drones to speed up survey work and protect sites from squatters, builders and miners. Small drones helped researchers produce three-dimensional models of Peruvian sites instead of the usual flat maps—and in days and weeks instead of months and years.

Drones have replaced expensive and clumsy small planes, kites and helium balloons. Drones costing as little as£650 have proven useful. In 2013 drones have flown over at least six Peruvian archaeological sites, including the colonial Andean town Machu Llacta 4,000 metres (13,000 ft) above sea level. The drones continue to have altitude problems in the Andes, leading to plans to make a drone blimp, employing open source software.

Jeffrey Quilter, an archaeologist with Harvard University said, "You can go up three metres and photograph a room, 300 metres and photograph a site, or you can go up 3,000 metres and photograph the entire valley."

Future potential

In the future, UAVs may be able to perform a variety of unique tasks apart from what they are capable of today. Engineers are currently working to produce remotely piloted UAVs that are capable of air to air combat, aerial refueling, combat search and rescue with facial recognition, and resupply to agents on the ground.

DESIGN AND DEVELOPMENT CONSIDERATIONS

UAV design and production is a global activity with manufacturers all across the world. The United States and Israel were initial pioneers in this technology, and U.S. manufacturers had a market share of over 60% in 2006, with U.S. market share due to increase by 5–10% through 2016.

Northrop Grumman and General Atomics are the dominant manufacturers in this industry on the strength of the Global Hawk and Predator/Mariner systems.

The FAA Modernization and Reform Act of 2012 required selection of six test sites from 24 applicant states, to study UAS safety and certify commercial UASs for use in U.S. airspace. Some universities offer UAS research and training programs or academic degrees.

According to the Stockholm International Peace Research Institute, Israeli companies were behind 41% of all UAVs exported in 2001-2011. The European market share represented 4% of global revenue in 2006.

Development costs for American military UAVs, as with most military programs, have tended to overrun their initial estimates. This is mostly due to changes in requirements during development and a failure to leverage UAV development programs over multiple armed services.

This has caused United States Navy UAV programs to increase in cost from 0% to 5%, while United States Air Force UAV programs have increased from 60% to 284%.

Degree of autonomy

Fig. UAV monitoring and control at CBP

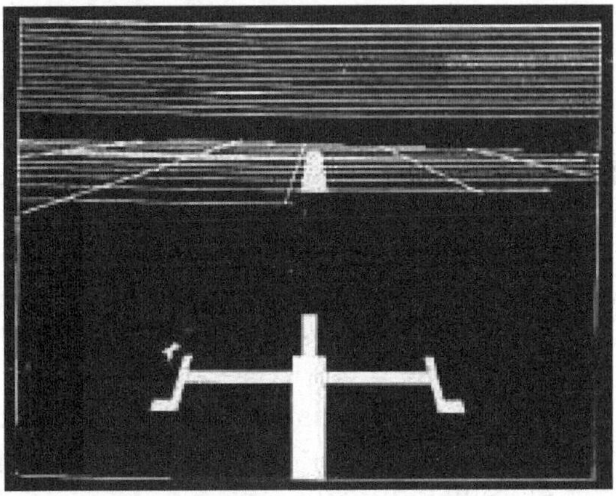

Fig. HiMAT Remote Cockpit Synthetic Vision Display (Photo: NASA 1984)

Early UAVs used during the Vietnam War captured video that was recorded to film or tape on the aircraft. These aircraft often flew either in a straight line or in preset circles collecting video until they ran out of fuel and landed. After landing, the film was recovered for analysis. Because of the simple, repetitive nature of these operations, the aircraft were often called "drones". As new radio control systems became available, UAVs were often remote controlled, and the term "remotely piloted vehicle" came into vogue. Today's UAVs often combine remote control and computerized automation. More sophisticated versions may have built-in control and/or guidance systems to perform low-level human pilot duties, such as speed and flight-path stabilization, and simple scripted navigation functions, such as waypoint following. In news and other discussions, the term "drone" is still often mistakenly used to refer to these more sophisticated aircraft.

From this perspective, most early UAVs were not autonomous at all. In fact, the field of air-vehicle autonomy is a recently emerging field, largely driven by the military to develop battle-ready technology. Compared to the manufacturing of UAV flight hardware, the market for autonomy technology is fairly immature and undeveloped.

Autonomy technology that is important to UAV development falls under the following categories:

- Sensor fusion: Combining information from different sensors for use on board the vehicle
- Communications: Handling communication and coordination between multiple agents in the presence of incomplete and imperfect information
- Path planning: Determining an optimal path for vehicle to follow while meeting certain objectives and mission constraints, such as obstacles or fuel requirements
- Trajectory Generation (sometimes called Motion planning):

Determining an optimal control maneuver to take in order to follow a given path or to go from one location to another

- Trajectory Regulation: The specific control strategies required to constrain a vehicle within some tolerance to a trajectory
- Task Allocation and Scheduling: Determining the optimal distribution of tasks amongst a group of agents within time and equipment constraints
- Cooperative Tactics: Formulating an optimal sequence and spatial distribution of activities between agents in order to maximize the chance of success in any given mission scenario

Autonomy is commonly defined as the ability to make decisions without human intervention. To that end, the goal of autonomy is to teach machines to be "smart" and act more like humans.

The keen observer may associate this with the developments in the field of artificial intelligence made popular in the 1980s and 1990s, such as expert systems,neural networks, machine learning, natural language processing, and vision. However, the mode of technological development in the field of autonomy has mostly followed a bottom-up approach, such as hierarchical control systems, and recent advances have been largely driven by the practitioners in the field of control science, not computer science. Similarly, autonomy has been and probably will continue to be considered an extension of the controls field.

To some extent, the ultimate goal in the development of autonomy technology is to replace the human pilot. It remains to be seen whether future developments of autonomy technology, the perception of the technology, and, most importantly, the political climate surrounding the use of such technology will limit the development and utility of autonomy for UAV applications. Also as a result of this, synthetic vision for piloting has not caught on in the UAV arena as it did with manned aircraft.

NASA utilized synthetic vision for test pilots on the HiMAT program in the early 1980s, but the advent of more autonomous UAV autopilots greatly reduced the need for this technology.

Interoperable UAV technologies became essential as systems proved their mettle in military operations, taking on tasks too challenging or dangerous for troops. NATO addressed the need for commonality through STANAG (Standardization Agreement) 4586. According to a NATO press release, the agreement began the ratification process in 1992. Its goal was to allow allied nations to easily share information obtained from unmanned aircraft through common ground control station technology. Aircraft that adhere to the STANAG 4586 protocol are equipped to translate information into standardized message formats; likewise, information received from other compliant aircraft can be transferred into vehicle-specific messaging formats for seamless interoperability. Amendments have since been made to the original agreement based on expert feedback from the field and an industry panel known as the Custodian Support Team. Edition Two of STANAG 4586 is currently under review. There are many systems available today that are developed in accordance with STANAG 4586, including products by industry leaders such as AAI Corporation, CDL Systems, and Raytheon, all three of which are members of the Custodian Support Team for this protocol.

Endurance

Because UAVs are not burdened with the physiological limitations of human pilots, they can be designed for maximized on-station times. The maximum flight duration of unmanned aerial vehicles varies widely. Internal combustion engine aircraft endurance depends strongly on the percentage of fuel burned as a fraction of total weight (theBreguet endurance equation) and so is largely independent of aircraft size.

Fig. UEL UAV-741 Wankel engine for UAV operations.

Because of the small size, weight, very low vibration and high power to weight ratio, Wankel rotary engines are increasingly being used in UAV aircraft. The engine is approximately one third of the size and weight of a piston engine of equivalent power output, which offers significant advantages for UAV aircraft.

Additionally: the engine rotors cannot seize, since rotor casings expand more than rotors; the engine is not susceptible to shock-cooling during descent; it does not require an enriched mixture for cooling at high power and having no reciprocating parts, there is less vulnerability to damage when the engines revolves higher than the designed maximum running operation. The attributes of the Wankel engine transpire into less fuel usage in UAVs giving greater range or a higher payload.

Solar-electric UAVs hold potential for unlimited flight, a concept originally championed by the AstroFlight Sunrise in 1974 and the much later Aerovironment Helios Prototype, which was destroyed in a 2003 crash.

Electric UAVs kept aloft indefinitely by laser power-beaming technology represent another proposed solution to the endurance challenge. This approach is advocated by Jordin Kare and Thomas Nugent.

One of the major problems with UAVs is the lack of inflight refueling capability. In 2012, the US Air Force was promoting research that should end in an inflight UAV refueling capability. A UAV-UAV simulated refuelling flight using two Global Hawks was achieved in 2012.

One of the uses for a high endurance UAV would be to "stare" at the battlefield for a long period of time to produce a record of events that could then be played backwards to track where improvised explosive devices (IEDs) came from. Air Force Chief of Staff John P. Jumper started a program to create these persistent UAVs, but this was stopped once he was replaced.

Notable high endurance flights			
UAV	**Flight time**	**Date**	**Notes**
QinetiQ Zephyr Solar Electric	336 hours 22 minutes	9–23 July 2010	
QinetiQ Zephyr Solar Electric	82 hours 37 minutes	28–31 July 2008	
Boeing Condor	58 hours 11 minutes	1989	The aircraft is currently in the Hiller Aviation Museum, CA.
Penguin B UAV Factory	54 hours 27 minutes	5–7 July 2012	
RQ-4 Global Hawk	33.1 hours	March 22, 2008	Set an endurance record for a full-scale, operational unmanned aircraft.
QinetiQ Zephyr Solar Electric	54 hours	September 2007	
IAI Heron	52 hours	?	
Israel Aerospace Industries Eitan	70+ hours	?	
AC Propulsion Solar Electric	48 hours 11 minutes	June 3, 2005	
MQ-1 Predator	40 hours 5 minutes	?	
GNAT-750	40 hours	1992	
TAM-5	38 hours 52 minutes	August 11, 2003	Smallest UAV to cross the Atlantic
Aerosonde	38 hours 48 minutes	May 3, 2006	
Shahed 129	24 hours	2012	
TAI Anka	24 hours	30 December 2010	
Vanguard Defense Industries	2 hours 55 minutes	11 February 2011	VTOL platform carrying an 18 lb payload.

In 2007, the Defense Advanced Research Projects Agency (DARPA) revealed a program to develop technology for a UAV with an endurance capability of over 5 years. The program, entitled VULTURE (Very-high altitude, Ultra-endurance, Loitering Theater Unmanned Reconnaissance Element), entered Phase II on September 14, 2010, with a contract signed with Boeing for development of the SolarEagle flight demonstrator.

Detect and avoid

The U.K.'s Civil Aviation Authority (CAA) has stated that it will require non-military drones larger than 20 kg to be able to automatically sense other aircraft and steer to avoid them, a technology still missing in civilian UAVs as of 2012.

Hardening of the control stations

Given the increasing military use of cyber attacks against Microsoft software, the United States Armed Forces have moved towards Linux ground control software.

Buddy attacks

The USAF said in 2012 that it will focus development of UAVs to be collaboratively networked with manned aircraft in "buddy attacks," while continuing to be able to fly as standalone systems.

EXISTING UAV SYSTEMS

UAVs have been developed and deployed by many countries around the world. For a list of models by country, see: List of unmanned aerial vehicles. The use of unmanned aerial systems, however, is not limited to state powers: non-state actors can also build, buy and operate these combat vehicles.

The export of UAVs or technology capable of carrying a 500 kg payload at least 300 km is restricted in many countries by the Missile Technology Control Regime. At the center of the American military's continued UAV research is the MQ-X, which builds upon the capabilities of the Reaper and Predator UAVs. As currently conceived, the MQ-X would be a stealthier and faster fighter-plane sized UAV capable of any number of missions: high-performance surveillance; attack options, including retractable cannons and bomb or missile payloads; and cargo capacity.

China has exhibited some UAV designs, but its ability to operate them is limited by the lack of high endurance domestic engines, satellite infrastructure, and operational experience.

Historical events involving UAVs

- In 1981, the Israeli IAI Scout drone, is operated in combat missions by the South African Defence Force against Angola during Operation Protea.
- In 1982, UAVs operated by the Israeli Air Force are instrumental

during Operation Mole Cricket 19, where both IAI Scout and Tadiran Mastiff are used to identity SAM sites, while Samson decoy UAVs are used to activate and confuse Syrian radar.

- During the Persian Gulf War, Iraqi Army forces surrendered to the UAVs of the USS Wisconsin.
- In October 2002, a few days before the U.S. Senate vote on the Authorization for Use of Military Force Against Iraq Resolution, about 75 senators were told in closed session that Saddam Hussein had the means of delivering biological and chemical weapons of mass destruction by UAVs that could be launched from ships off the Atlantic coast to attack U.S. eastern seaboard cities. Colin Powell suggested in his presentation to the United Nations that they had been transported out of Iraq and could be launched against the U.S. It was later revealed that Iraq's UAV fleet consisted of only a few outdated Czech training drones. At the time, there was a vigorous dispute within the intelligence community as to whether CIA's conclusions about Iraqi UAVs were accurate. The U.S. Air Force, the agency most familiar with UAVs, denied outright that Iraq possessed any offensive UAV capability.
- The first US targeted UAV killing outside the conventional battlefield took place on November 3, 2002, in the Marib district of Yemen. Six alleged terrorists were killed in their SUV by a UAV-fired missile. The command centre was in Tampa, Florida, USA.
- In December 2002, the first ever dogfight involving a UAV occurred when an Iraqi MiG-25 and a U.S. RQ-1 Predator fired missiles at each other. The MiG's missile destroyed the Predator.
- The U.S. deployed UAVs in Yemen to search for and kill Anwar al-Awlaki, an American and Yemen imam, firing at and failing to kill him at least once before he was killed in a UAV-launched missile attack in Yemen on 30 September 2011. The targeted killing of an American citizen was unprecedented. However, nearly nine years earlier in 2002, U.S. citizen Kemal Darwish was one of six men killed by the first UAV strike outside a war zone, in Yemen.
- In December 2011, Iran captured a United States' RQ-170 unmanned aerial vehicle that flew over Iran and rejected President Barack Obama's request to return it to the US. Iranian officials claim to have recovered data from the U.S. surveillance aircraft. However, it is not clear how Iran brought it down. There have also been claims that Iran spoofed the GPS signal used by the UAV and hijacked it into landing on an Iranian runway.

DOMESTIC AERIAL SURVEILLANCE

Although UAVs are today most commonly associated with military actions, UAVs are increasingly used by civilian government agencies, businesses, and private individuals. In the United States, for example, government agencies use UAVs such as the RQ-9 Reaper to patrol the nation's borders, scout property, and locate fugitives. One of the first authorized for domestic usage was the ShadowHawk UAV in service in Montgomery County, Texas, and is being used by their SWAT and emergency management offices.

UAVs over Australia

Sydney Harbour Bridge collision

On 2 October 2013 a UAV collided with Sydney Harbour Bridge. The craft, which carried a camera, was found about 10pm near a southern pylon of the bridge.Although it was found the day before the International Navy Fleet Review, police believed there was no connection with the event. The Civil Aviation Safety Authoritystarted an investigation and anti-terrorism officers were alerted, though police said they didn't believe it was suspicious and was for recreational use.

A CASA spokesman said that they had been contacted by the New South Wales Police Force. He added that those operating remotely piloted aircraft should keep them at least 30m away from structures, buildings and people, to check with the local council where they could be used and that the airspace around the bridge was restricted for all aircraft, including small ones.

UAVs over Belgium

In 2013, The Flemish Research Institute for Nature and Forest and the Flanders Marine Institute planned to use a UAV with a camera and a detection system that automatically recognizes different types of vegetation, trees and individual plants in order to make regular assessments of the biodiversity of the Flemish terrain.

Their previous system of using 25 civil servants to map the country was unsatisfactory and time-consuming.

UAVs over Democratic Republic of Congo

In August 2013, the Italian defence company Selex ES provided an unarmed surveillance drone to be deployed in the Democratic Republic of Congo to monitor movements of armed groups in the region and to protect the civilian population more effectively.

UAVs over Germany

The Deutsche Bahn (German national railways) said in 2013 that it would test small surveillance UAVs with thermal cameras to prosecute vandals who spray graffiti on its property at night. Graffiti incidents cost the Deutsche Bahn $10 million per year to clean up.

UAVs over Japan

T-Hawk and Global Hawk drones were used to gather information about the damaged Fukushima Number 1 nuclear plant and disaster-stricken areas of the Tohoku region after the March 2011 tsunami. Anti-whaling activists used an Osprey UAV (made by Kansas-based Hangar 18) in 2012 to monitor Japanese whaling ships in the Antarctic.

UAVs over Nepal

In 2012, the World Wildlife Fund supplied two FPV Raptor 1.6 UAVs to the Nepal National Parks. These UAVs were used to monitor rhinos tigers and elephants and deter poachers. The UAVs were equipped with time lapse cameras and could fly for 18 miles at 650 feet.

UAVs over the Republic of Ireland

In 2012 Loitering Theatre group flew Parrot AR.Drones near Áras an Uachtaráin, the Magazine Fort, a prison facility and offices of Google on Barrow Street and also Facebook. Footage filmed by the group was shown at an exhibit run by the Dublin Science Gallery. The Irish Aviation Authority stated that this was prohibited as Dublin city is classed as a restricted area.

A separate group, called Tomorrows Thoughts Today, who had attended the same event were detained on their return to the UK at London Southend Airport under theTerrorism Act. They were released after a couple of hours questioning.

Legal status

In May 2012 the Irish Aviation Authority published a document setting out safety requirements for any Unmanned Aerial System, regardless of mass. An appendix contained an application form to apply to operate a UAS. The only previous legislation had been the 'Irish Aviation Authority (Rockets and Small Aircraft) Order, 2000'.

The IAA policy is that Unmanned Aerial Systems may not be flown without the operator receiving a specific Permission from the IAA. Where such a craft is to be used for commercial work, the operator must apply to the IAA for an Aerial Work Permission from the IAA. Flying UAS outside the direct, unaided line of sight of the operator is not allowed for safety reasons. It is not permitted to use vision-enhancing systems, such as First Person View.

On 15 November 2012 the Irish Aviation Authority introduced a requirement that Remotely Piloted Aircraft needed to be registered to comply with Statutory Instrument 634 or 2005 'Nationality and Registration of Aircraft' Order.

UAVs over South Africa

In December 2012, the Kruger National Park started using a Seeker II UAV against rhino poachers. The UAV was loaned to the South African National Parks authority by its manufacturer, Denel Dynamics of South Africa.

UAVs over the United Kingdom

In 2007 Merseyside Police was reported to be conducting tests with a UAV. Merseyside Police caught a car thief with a UAV in 2010, but about a week later had to stop UAV operations as the craft was not licenced as new legislation required. A licence was granted by the Civil Aviation Authority. The UAS was lost during a training exercise in Aigburth, Liverpool, when crashing into the River Mersey. Merseyside Police stated the UAV would not be replaced, citing performance problems and cost of staff training.

In 2012, the Ulster Society for the Prevention of Cruelty to Animals used a quadcopter UAV to deter badger baiters in Northern Ireland. In March 2013, the British League Against Cruel Sports announced they had carried out trial flights with UAVs and planned to use a fixed-wing OpenRanger and an Octocopter to gather evidence to make private prosecutions against illegal hunting of foxes and other animals. The UAVs were supplied by ShadowView Ltd, Berkshire.

A spokesman for Privacy Internationalsaid that "licencing and permission for drones is only on the basis of health and safety, without considering whether privacy rights are violated." The Civil Aviation Authority (United Kingdom) rules are that UAV aircraft less than 20 kilograms in weight must be in direct visual contact with the pilot, cannot fly within 150 meters of a congested area or within 50 meters of a person or vehicle, and cannot be used for commercial activity. Staffordshire police were reported to be using UAVs as alternatives to standard police support units in 2012.

UAVs over the United States

Surveillance and policing

UAVs can be powerful surveillance tools by carrying camera systems capable of license plate scanning and thermal imaging as well as radio equipment and other sensors. The Electronic Frontier Foundation filed a Freedom of Information Act request on January 10, 2012 against the Federal Aviation Administration.

As a result of the request, the FAA released a list of the names of all public and private entities that have applied for authorizations to fly UAVs domestically. Some of these government licenses belong to the U.S. Customs and Border Protection, a component of the Department of Homeland Security. UAVs have been used by U.S. Customs and Border Protection to patrol United

States borders since 2005, and the agency currently owns 10 UAVs. With plans to use armed drones.

A May 2012 report issued by the DHS Inspector General found that CBP "needs to improve planning of its unmanned aircraft systems program to address its level of operation, program funding, and resource requirements, along with stakeholder needs". Also, despite the Bureau's limited mission to safeguard the borders, the Bureau often flies missions for the FBI, the Department of Defense, NOAA, local law enforcement, and other agencies. In December 2011, the CBP made headlines when reporters discovered that the agency's UAVs were being used to assist local law enforcement in relation to cattle raiding in North Dakota without receiving prior approval from the FAA or any other agency.

Individuals in the United States have few legal privacy protections from aerial surveillance conducted through UAVs. In Florida v. Riley, the United States Supreme Court held that individuals do not have the right to privacy from police observation from public airspace.

The weakness of legal protection from UAV surveillance have led to calls from civil liberties advocacy groups for the U.S. government to issue laws and regulations that establish both privacy protections and greater transparency regarding the use of UAVs to gather information about individuals. As an example, the ACLU has warned of a "nightmare scenario" in the future where the police might be able, with computer technology, to combine mobile phone tracking with video data and build up a database of people's routine daily movements.

In February 2013 Seattle mayor Michael McGinn ordered the Seattle Police Department to abandon plans to use UAVs after objections from residents. Two DraganflyerX6 craft had been purchased with a federal grant and the police had been granted FAA approval though they had not started using them. The vehicles were to be returned to the manufacturer. Seattle Police Department had announced in October 2012 that they were drafting a policy and they were one of the first police forces in the United States to receive approval from the federal government to use UAVs. Opponents of the programme included the Washington chapter of the American Civil Liberties Union.

On February 24, 2012, the Electronic Privacy Information Center, joined by over 100 organizations, experts, and members of the public, submitted a petition to the FAA requesting a public rule-making on the privacy impact of UAV use in U.S. airspace. In June 2012, Senator Rand Paul and Representative Austin Scott both introduced legislation that would require law enforcement to obtain a warrant before using a UAV to conduct surveillance of criminal activities. EPIC has stated that transparency and accountability must be built into the FAA's system of UAV/UAS/RPV regulation in order to provide basic protections to the public.

While Congress rapidly moves ahead to authorize further use of domestic UAVs, many remain skeptical regarding privacy concerns. Some privacy scholars argue that the domestic use of UAVs for surveillance will ultimately benefit privacy by encouraging society to demand greater privacy rights.

Associated today with the theatre of war, the widespread domestic use of drones for surveillance seems inevitable. Existing privacy law will not stand in its way. It may be tempting to conclude on this basis that drones will further erode our individual and collective privacy. Yet the opposite may happen. Drones may help restore our mental model of a privacy violation. They could be just the visceral jolt society needs to drag privacy law into the twenty-first century.

—M. Ryan Calo

FBI Director Robert Mueller testified before the Senate Judiciary Committee on the 19th of June 2013 that the FBI owns and uses UAVs for the purposes of "surveillance".

Non-police uses

Law enforcement and other government agencies are not the only entities that use UAVs. Private citizens and media organizations use UAVs as well for the purposes of surveillance, recreation, or personal land assessment. Some farming initiatives utilize UAVs for crop spraying, as they are often cheaper than a full-sized helicopter. Occupy Wall Street journalist Tim Pool uses what he calls an Occucopter for live feed coverage of Occupy movement events. The "occucopter" is an inexpensive radio controlledquadcopter with cameras attached and controllable by Android devices or iOS. In February 2012, an animal rights group used a MikroKopter hexacopter to film hunters shooting pigeons in South Carolina. The hunters then shot the UAV down. UAVs also have been shown to have many other civilian uses, such as agriculture, Hollywood, and in the construction industry.

JFK International incident

In March 2013 an Alitalia pilot on final approach to runway 31 right at John F. Kennedy International Airport reported seeing a small UAV near his aircraft.Both the FAA and FBI were reported to be investigating.

Virginia Bull Run crash

In August 2013 a UAV filming events at the Virginia Bull Run in Dinwiddie County, Virginia crashed into the crowd, causing minor injuries.

Manhattan drone crash

In September 2013 a UAV flying over Manhattan collided with a building and crashed on the pavement below near a businessman who reported the incident to the police.The pilot seemed to lack experience controlling it as it collided with several buildings before the collision that made it crash. In October 2013 it was reported that a man had been arrested days after the incident had been reported in the media and that he had been charged with reckless endangerment. He was identified because he was seen in the video recorded by the drone.

New York Drone Conference

In October 2013, the 'Drones and Aerial Robotics Conference' was held at New York University. The attendees included hobbyists and academics. A protest byGranny Peace Brigade against military and police use of drones was held outside.

UAVs over Venezuela

In June 2012, Venezuelan President Hugo Chavez claimed that Venezuela had begun producing its own UAV. General Julio Morales added that the UAV had a range of 100 kilometres (about 60 miles), a maximum altitude of 3,000 metres (about 10,000 feet), could fly for 90 minutes, measured three by four metres, and was a part of a system to survey and monitor pipelines, dams, and other rural infrastructure. General Morales was the president of the state-run Cavim arms manufacturer that developed the aircraft.

UAVs over Vietnam

In May 2013, The Vietnam Space Technology Institute successfully conducted 37 UAV flights in the central Lam Dong province. Research for the UAVs began in 2008 and was later funded by the state in 2011.

UAV OPERATORS

In the U.S., thousands of civilian UAV operators work for contractors, piloting and maintaining UAVs. A 2011 study by the Air Force School of Aerospace Medicineindicated that nearly 50% of spy UAV operators suffer from high stress-levels.

The president of a civilian UAV operators' union, the Association of Unmanned Operation (AUO), cited long working hours and decreasing wages as U.S. involvement in wars in Iraq and Afghanistan was reduced and as a result of the U.S. government's budget sequestration.

An August 2013 Brookings Institution study reported that there were approximately 1,300 remotely piloted aircraft (RPA) pilots in the U.S. Air Force, 8.5 percent of total Air Force pilots, up from 3.3 percent in 2008. The study indicated that the U.S. military's combat air patrol (CAP) daily missions requirement is growing at a faster pace than RPA pilots can be trained, with an attrition rate during RPA flight screening being three times that of traditional pilots and a 13% lower promotion rate to Major than other officers.

UAVS IN POPULAR CULTURE

- UAVs have been used in many episodes of two recent science-fiction television series, Stargate SG-1 and Dark Angel.
- A UCAV AI, called EDI, was central to the sci-fi action film Stealth.
- UAVs are also used in video games, such as F.E.A.R., inFamous, Tom Clancy's Ghost Recon, and the popular Call of Duty and Battlefield franchises.
- An MQ-9 reaper controlled by a rogue supercomputer appears in the film Eagle Eye (2008).
- The hapless would-be terrorists in the film Four Lions (2010) are targeted by and attempt to shoot down an RQ-1 Predator.
- The Bourne Legacy (2012 film) features a Predator UAV pursuing the protagonists.
- An episode of the TV show Castle, first broadcast in May 2013, featured a UAV hacked by terrorists.

PUBLIC OPINION

In February 2013, Fairleigh Dickinson University's PublicMind poll found that 48% of American voters believe it is "illegal for the U.S. government to target its own citizens living abroad with drone attacks. Just 24% say that it's legal." "The public clearly makes an assumption very different from that of the Obama administration or Mr. Brennan: the public thinks targeting American citizens abroad is out of bounds," Peter Woolley, founding director of PublicMind and professor of political science at FDU, said to CNN.

In the same poll, however, by a wide six-to-one margin (75% vs.13%), voters approved of the U.S. military using UAVs to carry out attacks abroad "on people and other targets deemed a threat to the U.S." Republicans, men, and whites approve more strongly than Democrats, women, and non-whites, but approval is robust in all demographic categories. Voters also approve of the CIA using UAVs to carry out attacks abroad by a strong three-to-one margin (65% vs. 21%), but this approval is significantly less than approval for the U.S. military carrying out such attacks.

Despite this broad-based public support, there are a number of vocal critics of the increasing use of UAVs to track and kill terrorists and militants. A major criticism of drone strikes is that they result in excessive collateral damage. David Kilcullen and Andrew Exum wrote in the New York Times that "according to Pakistani sources, they have also killed some 700 civilians. This is 50 civilians for every militant killed, a hit rate of 2 percent." Other studies have put the civilian casualty rate anywhere between 4 and 35 percent. It is difficult to reconcile these figures because the drone strikes are often in areas that are inaccessible to independent observers and the data includes reports by local officials and local media, neither of whom are reliable sources.

Critics also fear that by making killing seem clean and safe, so-called surgical UAV strikes will allow the United States to remain in a perpetual state of war. However, others maintain that drones "allow for a much closer review and much more selective targeting process than do other instruments of warfare" and are subject to Congressional oversight. Like any military technology, armed UAVs will kill people, combatants and innocents alike, thus "the main turning point concerns the question of whether we should go to war at all."

Morality

The "unmanned" aspect of UAVs is primarily what sets them apart from manned aircraft. This aspect also raises certain moral concerns. Some believe that the asymmetry of fighting humans with machines that are controlled from a safe distance lacks integrity and honor that was once valued during warfare. Others feel that if such technology is available, then there is a moral duty to employ it in order to save as many lives as possible. Another potential moral issue with UAVs is that because they do not allow for pilot casualties, some fear that they will be used more frivolously, and that human lives affected by UAV-based strikes will not be regarded with as much consideration as with manned aerial attacks.

Legality

A confidential Justice Department memo concludes that the U.S. government can order the killing of American citizens if they are believed to be "senior operational leaders" ofAl-Qaeda or "an associated force"—even if there is no intelligence indicating they are engaged in an active plot to attack the U.S. The secrecy surrounding such strikes is quickly emerging as a central issue in the hearing of White House counterterrorism adviser John O. Brennan, a key architect of the UAV campaign, to be CIA director.

Brennan was the first administration official to publicly acknowledge UAV-based strikes in a speech last year, calling them "consistent with the inherent right of self-defense." In a separate talk at the Northwestern University Law School, Attorney General Eric Holder specifically endorsed the constitutionality of targeted killings of Americans, saying they could be justified if government officials determine that the target poses "an imminent threat of violent attack." But the confidential Justice Department "white paper" introduces a more expansive definition of self-defense or imminent attack than described by Brennan or Holder in their public speeches It refers, for example, to what it calls a "broader concept of imminence" than actual intelligence about any ongoing plot against the U.S. homeland.

"The condition that an operational leader presents an 'imminent' threat of violent attack against the United States does not require the United States to have clear evidence that a specific attack on U.S. persons and interests will take place in the immediate future," the memo states. Instead, an "informed, high-level" official of the U.S. government may determine that the targeted American has been "recently" involved in "activities" posing a threat of a violent attack and "there is no evidence suggesting that he has renounced or abandoned such activities." (The memo does not define "recently" or "activities.") As in Holder's speech, the confidential memo lays out a three-part test that would make targeted killings of American lawful: in addition to the suspect being an imminent threat, capture of the target must be"infeasible, and the strike must be conducted according to "laws of war principles."

But the memo elaborates on some of these factors in ways that go beyond what the attorney general said publicly. For example, it states that U.S. officials may consider whether an attempted capture of a suspect would pose an "undue risk" to U.S. personnel involved in such an operation. If so, U.S. officials could determine that the capture operation of the targeted American would not be feasible, making it lawful for the U.S. government to order a killing instead, the memo concludes.

The undated memo is entitled "Lawfulness of a Lethal Operation Directed Against a U.S. Citizen who is a Senior Operational Leader of Al Qaeda or An Associated Force." It was provided to members of the Senate Intelligence and Judiciary committees in June 2013 by administration officials on the condition that it be kept confidential and not discussed publicly. Although not an official legal memo, the white paper was represented by administration officials as a policy document that closely mirrors the arguments of classified memos on targeted killings by the Justice Department's Office of Legal Counsel, which provides authoritative legal advice to the president and all executive branch agencies.

The administration has refused to turn over to Congress or release those memos publicly or to even publicly confirm their existence. A source with access to the white paper, which is not classified, provided a copy to NBC News. The white paper also includes a more extensive discussion of why targeted strikes against Americans does not violate constitutional protections afforded to American citizens as well as a U.S. law that criminalizes the killing of U.S. nationals overseas.

It also discusses why such targeted killings would not be a war crime or violate a U.S. executive order banning assassinations. "A lawful killing in self-defense is not an assassination," the white paper reads. "In the Department's view, a lethal operation conducted against a U.S. citizen whose conduct poses an imminent threat of violent attack against the United States would be a legitimate act of national self-defense that would not violate the assassination ban. Similarly, the use of lethal force, consistent with the laws of war, against an individual who is a legitimate military target would be lawful and would not violate the assassination ban."

UNMANNED COMBAT AIR VEHICLE

Fig. A MQ-9 Reaper during a training mission

An unmanned combat air vehicle (UCAV), also known as a combat drone or drone, is an unmanned aerial vehicle (UAV) that is usually armed. These aircraft have no onboard human pilot. Operational drones are predominantly under real-time human control, with "the human's role in UCAV system [varying] according to levels of autonomy of UCAV and data communication requirement[s]".

Equipment necessary for a human pilot (such as the cockpit, armor, ejection seat, flight controls, and environmental controls for pressure and oxygen) are not needed in a UAV, as the operator runs the vehicle from a remote terminal, resulting in a lower weight and size than a manned aircraft. This can allow for greater payloads, range and maneuverability, albeit at the expense of situational awareness.

HISTORY[EDIT]

One of the earliest explorations of the concept of the combat drone was by Dr. Lee De Forest, an early inventor of radio devices, and U. A. Sanabria, a TV engineer. They presented their idea in an article in a 1940 publication of Popular Mechanics. The modern military drone as known today was the brainchild of John Stuart Foster Jr., a nuclear physicist and former sex head of the Lawrence Livermore National Laboratory (then called the Lawrence Radiation Laboratory). In 1971, Foster was a model airplane hobbyist and had the idea this hobby could be applied to building weapons. He drew up plans and by 1973 Defense Advanced Research Projects Agency (DARPA) built two prototypes called "Praeire" and "Calere", they were powered by a modified lawn-mower engine and could stay aloft for two hours while carrying 28-pounds of load.

In the late 1980s, Iran deployed a drone armed with six RPG-7 rounds in the Iran–Iraq War. This was the first time an armed drone was used in war.

In the 1973 Yom Kippur War, Israel used U.S. Ryan Firebee target drones to spur Egypt into firing its entire arsenal of anti-aircraft missiles. This mission was accomplished with no injuries to Israeli pilots, who soon exploited the depleted Egyptian defenses. In the late 1970s and 80s, Israel developed the Scout and the Pioneer, which represented a shift toward the lighter, glider-type model of UAV in use today.

Israel pioneered the use of UAVs for real-time surveillance, electronic warfare, and decoys. The images and radar decoying provided by these UAVs helped Israel to completely neutralize the Syrian air defenses in Operation Mole Cricket 19 at the start of the 1982 Lebanon War, resulting in no pilots downed. And in 1987, Israel drones were used as proof-of-concept of super-agility [post-stall controlled flight] in combat flight simulations was with tailless, Stealth-Technology-based three-dimensional Thrust Vectoring flight control [jet steering] for the first time.

Impressed by Israel's success, the US quickly acquired a number of UAVs, and its Hunter and Pioneer systems are direct derivatives of Israeli models. The first 'UAV war' was the first Gulf War: according to a May 1991 Department of the Navy report: "At least one UAV was airborne at all times during Desert Storm." After the Gulf War successfully demonstrated their utility, global militaries invested widely in the domestic development of combat UAVs.

In recent years the U.S. has increased its use of drone strikes in Pakistan as part of the War on Terrorism.

LAWS AND ETHICS OF WAR

The international laws of war (such as the Geneva Conventions) govern the conduct of participants in war (and also define combatants). These laws place a burden upon participants to limit civilian deaths and injuries through proper identification of targets and distinction between combatants and non-combatants. The use of completely autonomous weapon systems is problematic, however, because of the difficulty in assigning accountability to a person. Therefore, current designs still incorporate an element of human control (a "man in the loop") – meaning that a ground controller must authorize weapons release.

Concerns also include the human controller's role, because if he is a civilian and not a member of the military (which is quite possible with developmental and highly sophisticated weapons systems) he would be considered a combatant under international law which carries a distinct set of responsibilities and consequences. It is for this reason that the "man in the loop" should ideally be a member of the military that understands and accepts his role as combatant.

Controllers can also experience psychological stress from the combat they are involved in. They may communicate with the ground troops they are supporting and feel a bond with them. They may also feel helplessness, guilt, exhaustion, or burnout as a response to what they witness remotely. A few may even experienceposttraumatic stress disorder (PTSD).

Professor Shannon E. French, the director of the Center for Ethics and Excellence at Case Western Reserve University and a former professor at the U.S. Naval Academy, wonders if the PTSD may be rooted in a suspicion that something else was at stake. According to Professor French, the author of the 2003 book The Code of the Warrior(ISBN 0847697568):

"If [I'm] in the field risking and taking a life, there's a sense that I'm putting skin in the game," she says. "I'm taking a risk so it feels more honorable. Someone who kills at a distance—it can make them doubt. Am I truly honorable?"

On 28 October 2009, United Nations Special Rapporteur on extrajudicial, summary or arbitrary executions, Philip Alston, presented a report to the Third Committee (social, humanitarian and cultural) of the General Assembly arguing that the use of unmanned combat air vehicles for targeted killings should be regarded as a breach ofinternational law unless the United States can demonstrate appropriate precautions and accountability mechanisms are in place.

The Missile Technology Control Regime applies to UCAVs.

Collateral damage of civilians still takes place with drone combat, although some (like John O. Brennan) have argued that it greatly reduces the likelihood. Although drones enable advance tactical surveillance and up-to-the-minute data, flaws can become apparent. The U.S. drone program in Pakistan has killed several dozen civilians accidentally for example. Another example is the operation in 2010 Feb near Khod, in Urozgan Province, Afghanistan. Over ten civilians in a three-vehicle convoy travelling from Daykundi Province were accidentally killed after a drone crew misidentified the civilians as hostile threats. A force of Bell OH-58 Kiowa helicopters, who were attempting to protect ground troops fighting several km away, fired AGM-114 Hellfire missiles at the vehicles.

As of March 2013, the evolution of laws governing drones use continued to be debated.

POLITICAL EFFECTS

As a new weapon, drones are having unforeseen political effects. Some scholars have argued that the extensive use of drones will undermine the popular legitimacy of local governments, which are blamed for permitting the strikes. The case study for this analysis is Yemen, where drone strikes seem to be increasing resentment against the Yemeni government as well as against the U.S.

Some leaders worry about the effect drone warfare will have on soldiers' psychology. Keith Shurtleff, an army chaplain at Fort Jackson, South Carolina, worries "that as war becomes safer and easier, as soldiers are removed from the horrors of war and see the enemy not as humans but as blips on a screen, there is very real danger of losing the deterrent that such horrors provide". Similar worries surfaced when "smart" bombs began to be used extensively in the First Gulf War.

Mark Bowden has refuted this viewpoint saying in his The Atlantic article, "But flying a drone, [the pilot] sees the carnage close-up, in real time—the blood and severed body parts, the arrival of emergency responders, the anguish of friends and family. Often he's been watching the people he kills for a long time before pulling the trigger. Drone pilots become familiar with their victims. They see them in the ordinary rhythms of their lives—with their wives and friends, with their children. War by remote control turns out to be intimate and disturbing. Pilots are sometimes shaken."

Cultural and career issues have lead to a shortfall in USAF drone operators, which is seen as a "dead end job".

FUTURE MODELS

Fig. A BAE Raven during flight testing

Fig. The EADS Barracuda on the ManchingAir Base in Germany

Overview

Note: Some of these are not aircraft prototypes but technology demonstrators (TD) that are not expected to enter service.

Name	Manufacturer(s)	Developing nation(s)
A-10PCAS	Elbit Systems	Israel
Sky-X (TD)	Alenia Aeronautica	
Eitan	Israel Aerospace Industries	Israel
Harop	Israel Aerospace Industries	Israel
Burraq	... Pakistan Air Force ... National Engineering and Scientific Commission	Pakistan
AURA	Defence Research and Development Organisation	India
Taranis (TD)	BAE Systems	United Kingdom
Dassault nEUROn (TD)	Dassault Aviation	... France ... Spain ... Italy ... Sweden ... Greece
Rustom (TD)	Defence Research and Development Organisation	India
Avenger	General Atomics Aeronautical Systems	United States
X-47A (TD)/B (TD)/C	Northrop Grumman	United States
Armstechno Dulo (TD)	Armstechno	Bulgaria
Bateleur (TD)	Denel Dynamics	South Africa
Anka	Turkish Aerospace Industries	Turkey
Ra'd	Armed Forces of the Islamic Republic of Iran	Iran
Karrar	Iran	Iran
Sofreh Mahi	Iran Aircraft Manufacturing Industrial Company	Iran
Shahed 129	Shahed Aviation Industries Research Center	Iran
Barracuda (TD)	EADS	... Germany ... Spain
Skat	Mikoyan	Russia

Various Chinese UCAV concepts have also materialized. WZ-2000, UCAV versions of the Xianglong high altitude are long endurance UAV. Also, dedicated UCAV's Shenyang's Dark Sword (Anjian), and also revealed at Zhuhai 2008 was a model of a stealth strike UCAV with forward swept wings, filling a similar niche to U.S. X-45 called the Warrior Eagle.

BAE Taranis

Taranis is a British demonstrator programme for unmanned combat air vehicle (UCAV) technology. It is part of the UK's Strategic Unmanned Air Vehicle (Experimental) programme (SUAV[E]). BAE describes Taranis's role in this context as following: "This £124m four year programme is part of the UK Government's Strategic Unmanned Air Vehicle Experiment (SUAVE) and will result in a UCAV demonstrator with fully integrated autonomous systems and low observable features." The Taranis demonstrator will have an MTOW (Maximum Takeoff Weight) of about 8000 kilograms and be of comparable size to the BAE Hawk – making it one of the world's largest UAVs.

Fig. BAE Taranis model, one of the largest design concepts

It will be stealthy, fast, and able to deploy a range of munitions over a number of targets, as well as being capable of defending itself against manned and other unmanned enemy aircraft. The first steel was cut in September 2007 and ground testing started in early 2009.

The first flight of the Taranis is planned for the first quarter of 2013. The demonstrator will have two internal weapons bays. With the inclusion of "full autonomy" the intention is thus for this platform to be able to "think for itself" for a large part of the mission.

J-UCAS

Fig. J-UCAS Boeing X-45A UCAV technology demonstrator

- Boeing X-45 UCAV (TD)
- Northrop-Grumman X-47 Pegasus

Joint Unmanned Combat Air Systems, or J-UCAS, was the name for the joint U.S. Navy/U.S. Air Force unmanned combat air vehicle procurement project. J-UCAS was managed by DARPA, the Defense Advanced Research Projects Agency.

In the 2006Quadrennial Defense Review, the J-UCAS program was terminated. The program would have used stealth technologies and allowed UCAVs to be armed with precision-guided weapons such as Joint Direct Attack Munition (JDAM) or precision miniature munitions, such as the Small-Diameter Bomb, which are used to suppress enemy air defenses. Controllers could have used real-time data sources, including satellites, to plan for and respond to changes on and around the battlefield.

The program was later revitalized into UCAS-D, a United States Navy program designed to develop a carrier-based unmanned aircraft.

N-UCAS

UCAS-D and Northrop Grumman X-47B are the U.S. Navy-only successors to the J-UCAS, which was cancelled in 2006. Boeing is also working on the X-45N in this sector.

In a New Year 2011 editorial titled "China's Naval Ambitions", The New York Times said "[t]he Pentagon must accelerate efforts to make American naval forces in Asia less vulnerable to Chinese missile threats by giving them the means to project their deterrent power from further offshore. Cutting back purchases of the Navy's DDG-1000 destroyer (with its deficient missile defense system) was a first step. A bigger one would be to reduce the Navy's reliance on short-range manned strike aircraft like the F-18and the F-35, in favor of the carrier-launched N-UCAS...."

On 6 January 2011, the DOD announced that this would be one area of additional investment in the 2012 budget request.

USAF Hunter-Killer

- Scaled Composites Model 395
- Scaled Composites Model 396
- General Atomics MQ-9 Reaper (originally the Predator B)
- Aurora Flight Sciences/Israel Aircraft Industries Eagle/Heron 2
- Unnamed Lockheed Martin entry

The United States Air Force has shifted its UCAV program from medium-range tactical strike aircraft to long-range strategic bombers. The technology of the Long Range Strike program is based on the Lockheed Martin Polecat demonstrator.

Sagem Sperwer

The Sagem Sperwer B is a long endurance tactical UAV capable of surveillance and armed combat missions. The Sperwer B improves on its predecessor, the Sperwer A, with twice the payload capacity and twice the endurance, offering a 100 kg payload and 12 hours of sustained flight.

Fig. Sagem Sperwer B (not weaponized on this photo)

It can be outfitted with electro-optic/infrared sensors, electronic and communications intelligence (ELINT/COMINT), synthetic aperture radar (SAR), as well as weapons payloads such as the Rafael Spike-LR anti-tank missile and the Bonus munition from Nexter/BAE Systems Bofors. All ground facilities of the Sperwer A (used by France, Netherlands, Sweden, Greece, Canada and Denmark) are compatible with the Sperwer B.

Elbit Hermes 450

The Israeli Air Force, which operates a squadron of Hermes 450s out of Palmachim Airbase south of Tel Aviv, has adapted the Hermes 450 for use as an assault UAV, reportedly equipping it with two Hellfire missiles or, according to various sources, twoRafael-made missiles. According to Israeli, Palestinian, Lebanese and independent reports, the Israeli assault UAV has seen extensive service in the Gaza Strip and was used intensively in the Second Lebanon War. Israel has not denied this capability, but to date, its policy has been not to officially confirm it either.

Lethal Miniature Aerial Munition System (LMAMS)

The LMAMS is a man-carried missile with many UAV characteristics such as loitering time and a hand-held ground station.

F-16

In 2013 Boeing retrofitted retired multiple Lockheed Martin F-16s (re-designated QF-16) with UAV technology. The company demonstrated combat maneuvers includingbarrel rolls and a "split S" (where the aircraft turns upside down and flies a descending half-loop, reversing direction) and a perfect center line landing. During the test flight, the plane cruised at 40,000 feet (12,000 m) and reached speeds of Mach 1.47. The aircraft reached 7Gs of acceleration, but was expected to operate at 9Gs. The existing six QF-16s will be augmented in 2015.

PUBLIC OPINION SUMMARY

In February 2013, Fairleigh Dickinson University's PublicMind poll conducted a study to measure public opinion on the use of drones. The study was conducted nationwide, and it asked registered voters whether they "approve or disapprove of the U.S. Military using drones to carry out attacks abroad on people and other targets deemed a threat to the U.S.?" The results showed that three in every four (75%) of voters approved of the U.S. Military using drones to carry out attacks, while (13%) disapproved.

THE RISE OF THE ROBOT POLICE

This chapter provides a glimpse at the international growth of police, surveillance and homeland security robots. Several countries are included here but it is by no means an exhaustive listing – there are more coming online every week.

[SH]USA

An increasing number of US police forces are now using robots, such as the ANDROS Remotec, for these tasks. Originally intended for bomb disposal, there are now daily reports of police and SWAT (Special Weapons And Tactics) teams using robots for a greater variety of dangerous tasks such as breaking down doors, approaching hostage takers, searching buildings, smashing windows and taking mobile phones to barricaded suspects.

Many of these reports have heart warming outcomes such as hostage children being safely released. Often the gunmen will just give themselves up when confronted by a robot even though it is not armed or dangerous. When children's lives are being saved in this way we feel a warm glow towards the robot and its handler. The trust such incidents create accumulate to make the application of police robots acceptable despite ethical "what if?" factors.

In June, 2007 the iRobot corp., makers of the PackBot for the US army and TASER international, the "stun gun" company announced a strategic alliance. As a first taster, the two companies have integrated the PackBot robot with an onboard TASER X26. This is a "non-lethal" weapon that uses compressed nitrogen to shoot darts attached to insulated wires into a human up to 10.6 meters away and deliver a blast of high voltage (up to 50,000 volts). The target's muscles contract uncontrollably allowing them to be dealt with either in a more conventional way or by having more shocks sent into their muscles.

This is just the beginning of arming police robots to enforce laws in the civilian world. Foster-Miller of Massachusetts also has an optional taser mounted on one of their Talon robots.

More alarmingly they have lethal weapon capability on their Talon SWAT robot (pictured) designed for police service. The robot can be configured with a 40 mm grenade launcher, a 12-gauge shotgun or a FN303 less-lethal launcher.

For now, police robots operate only under human control but, following on from the military, this is set to change. As I suggested earlier, once technology has been tried and tested in the theatre of war it will come home.

George Pike, the director of Global Security, talking about robot police, recently said that, "We may see autonomous robots capable of inflicting lethal force". He expects these toward the end of the next decade.

[SH]South Korea

In 2007, South Korea commissioned Samsung to build autonomous border guards. These can detect intruders in the demilitarized zone between North and South Korea at a 4Km range and automatically shoot them. According to the Korean Times, the South Korean Ministry of Information and Defence put $33.9 million USD into the development of police robot in 2006 with the purpose of developing a robot police force by 2101.

The Seoul authorities have also had a school guard robot developed by KT Telcop) to watch out for potential paedophiles in school playgrounds. The OFRO can autonomously patrol areas on pre-programmed routes. It is equipped with a camera system and microphone so that teachers can see what it can see.

The same robot was also used to to patrol the grounds for the World Cup in 2006 (pictured).

[SH]China

The Chinese have recently been developing police robots for use at the Beijing Olympics in 2008 including the Dragon Guard X3 portable robot developed by Shanghai's Grandar Robotics and the 56 kg robot, RAPTOR (pictured) made by Beijing Universal Pioneering Technology Co., Ltd. According to the People's Daily, China recently unveiled an armed robot for use at the Olympic Games.

Perhaps a more interesting glimpse of the futures is the Chinese Police Humanoid robot (pictured left) with four video cameras: three in the head and a pinhole camera in its chest. Citizens can press the red button on the robot's stomach and be automatically connected with police headquarters, and speak directly to an officer using a microphone installed on the robot's chest. According to Changping police, the robot will be used where cameras can't be installed and where incidents happen frequently.

Japan

This is the most advanced robot nation with a population that have a very positive attitude to them. They have a range of advanced security robots like the Reborg-Q (pictured) that communicate with and use lifts. It has been patrolling shopping centres in Japan to monitor unauthorised access as well as the presence of water leaks and fires. The machine has a touch-screen interface on its chest for visitors to get information about weather, the time and even missing kids. Now the company, Sohgo Security Servies, have deployed one in an office complex to keep track of employees' overtime working hours, gathering information on and reporting them to the HR department.

The Japanese also have a female police robot (pictured) that does not carry a gun, but wears the same cap as a female cop. It was introduced at Kokura Minami police station of Kitakyushu city in Japan. Local police authorities held an event at a mall to educate kids against street crime. Recently the Tokyo police used a sewer cleaning robot to hunt down a purse snatcher who had escaped into the pipework. It had a camera and spotlight so that it could be navigated.

UK

Bomb disposal robots have been used by the police for many years in the UK throughout the IRA period and they are not a novelty here. But more recently the police in Liverpool and Glasgow acquired German made hicam microdrones (pictured) for surveillance operations. The Staffordshire police deployed one at the 2007 V music festival. This is a little battery-powered helicopter weighing less than 1kg and is less than 1 meter in diameter. Combined with very quiet operation, the microdrone is equipped with video camera and a choice of sensing equipment that allows entirely covert operations. According to the suppliers, MS Power Systems, it can be used to track criminals and record anti-social behaviour as well as crowd and incident monitoring. Despite the reluctance of the US to follow our lead with surveillance cameras, the US police authorities are beginning to use unmanned aircraft for monitoring the population.

Russia

The Moscow State Technical University have developed a police robot that can record offences against the law using its five cameras, and also urge citizens to heed the law by reciting local statutes and repeating them over and over in a monotone voice. It weighs 250 kilograms, with a height of 180cm. It is being used to patrol streets in the city of Perm to tell drinkers to take their alcohol indoors. The robot has a panic button for passersby to call the police.

Israel

Israel are using a number of police robots that have been commissioned by the military because of their particular problems. Examples are the armed Viper robots and the more recent Guardium (pictured) which has some autonomous functioning.

Made by G-Nius Unmanned Ground Systems, it is operated from a command room, and can be mounted with cameras, night vision equipment and sensors, and machine guns. It can follow pre-programmed routes to navigate alone through cities. It can be used to patrol borders, scanning 360degrees and alerting operators of anything suspicious. It can travel at 80 km/h, and carry 660 pounds. It is controlled by means of a console that has a similar appearance to a video game controller.

TIMELINE FOR THE FUTURE OF ROBOT POLICING

This timeline is solidly based on 2008 technologies and discusses how they may converge to make future robots. It does not account for unknown technologies; nor does it take into account the technological response of criminals with jamming equipment, electronic cloaking and anti-robot weapons not to mention burglar robots. Such technology could push evolution of the police robot at a faster pace with more robust response. The timeline is based on predictions about the technological development and what will be available. Whether or not it will be deployed in the manner suggested here will depend on many complex societal and legislative factors that are beyond the remit of the report.

2008-2010

Wheeled robots will be used for disposing of explosives and mobile surveillance camera robots with loudspeaker and built in functions will continue to be used for monitoring potential criminal activity. There will be a progressive use of robots for security of buildings, airports and perimeter fences under control of operators; some aerial surveillance robots – micro-helicopters and fixed wing planes; some tracked robots with non-lethal weapons such as Tasers, pepper spray and water cannon and some with lethal weapons for flushing out terrorists and dangerous armed criminals.

There will be some semi-autonomous operation for specific functions like navigation, avoidance and escape but mostly the robots will be operated by games controllers and always supervised by humans. They will work alongside police operators much like a police dogs. Most of these developments are well under way.

2025

One of the most important extensions to the police use of robots in this period will be advanced biometric ID checks such as face, retina and fingerprint recognition. These facilities will greatly extend the surveillance potential of the robots and make their use considerably more widespread They will be able to determine the identity of individuals in seconds through direct link with police computers.

Wheeled humanoid police robots with more semi-autonomous operation will be in use on the streets. Many robots could sit passively in areas in town centres or areas known for crime. They could monitor any sudden scene changes like the appearance of a group or a crowd and use audio and vision systems to determine whether they are a group of harmless drunks or are potentially dangerous. Audio levels and speech perception would be used to tell if there was aggressive shouting and swearing and vision systems could watch for human contact or standoffs.

Robots will alert a human operator to survey the scene and advise. The robot could be remote controlled to approach the situation and enable the operator to ask questions, check IDs and assess the situation. Robot backup could be sent in with aerial vehicles and more ground robots.

2040

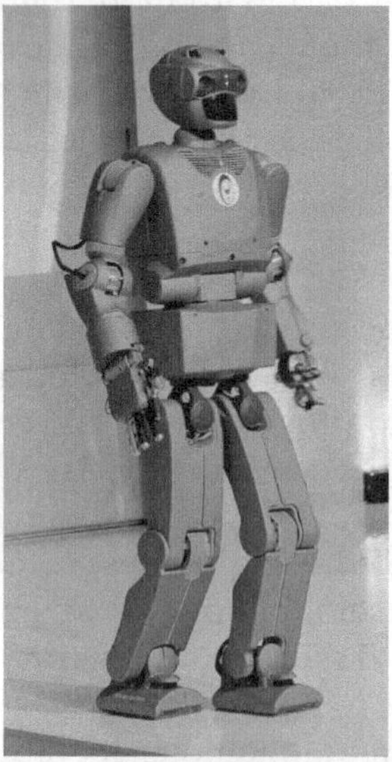

There will be much more autonomous functioning to reduce costs and free police officers for more human tasks. Humanoid walking robots would be more in use for crowd control at games, strikes and riots. Robots will patrol city centres and trouble spots where fights are likely to break out. Robots will have reasonable speech perception and be able to ask questions and respond to answers. What is your ID number? What are you doing here? Move along. They may work in teams of tracked robots with non-lethal weapons (e.g. Tasers or nets) and be on call for diffusing difficult situations and arresting people.

Powerful soft bodied robots will be developed that can restrain people without danger of hurting them. These could then be used as robot bouncers and security guards at nightclubs. There could also be radio tickets so robots can tell if humans have tickets and eject or detain those without. Robots will be able to spray a crowd with RFID tag darts or some futuristic equivalent so that people can be tracked after the crowd has been dispersed. They will always have a human operator on call to assist with ambiguities and to give instructions about the use of physical force.

2055

There will be more extensive use of robots with facial expressions and body gestures for dealing with people and diffusing difficult situations. The aim of these will be to build up human trust and confidence in the machine. Public acceptance will be required to legislate for more extensive robot application. such as autonomous robot traffic wardens. There will be absolutely no point in arguing with a robot traffic warden.

One big development in this period, with the greatest impact on personal liberty and privacy, will be the use of Extended Sensing by robots. Robots, unlike animals, are not limited to sensing within the confines of their own bodies. It makes no difference to a robot if its sensors and cameras are attached to its head to its feet or on the wall next to it. The control systems for factory robots and soccer robots often operate with cameras on the ceiling looking down on robot and environment to plan movements. A ground robot could have camera eyes and sensing flying overhead on micro-helicopters without any difficulty. In the same way, robots connected into a network of surveillance cameras are no different than a robot with multiple distributed eyes. Thus a single robot would be able to track an individual's movements throughout a city.

2070

The first autonomous police cars will have appeared – they have no need for a driver. These will be able to stop violators or use their cameras to identify the licence plate and automatically deduct fines from bank accounts. There are already a number of autonomous cars in 2008 but it is not possible to get them in operation on the roads because of insurance and legislative difficulties. Many competitive companies, not to mention the military, are working hard to overcome all of the difficulties and that should happen by this period.

Robots will use ultra high speed networks to communicate with each other as if they were a single machine so that they can act in concert. Each robot will be able to "see" what all of the others can see. Their computers will be just another node on the network so that information sharing is almost instantaneous. They will be able to search integrated internet databases at very high speed and access visual information from networks of survellience cameras or even access other robot visual systems and data banks.

Another development will be the use of swarm intelligence technologies either with small or large robots to make escape from capture impossible. When set in automatic mode, the robots will be able to collectively plan at very high speeds and communicate directly through their computers without giving anything away to suspects.

2084

As well as greater development of the robots that we have seen up until now, we can expect advanced humanoid robots made from squidgy inorganic material with human like features and expressions that make them appear almost human. They will not get angry or frustrated and could be made to always look friendly even when being shot at. Such inorganic "squidgy" android police will be the public face of police robotics to engender trust with the population. They will converse with the public to gather information, assess potentially dangerous situation and question suspects. They will disperse crowds and restrain drunks and 'see' with a network of surveillance cameras and unmanned micro aerial vehicles able to track and monitor throughout the future city.

Their biggest advantages will be speed of process, being able to run faster than humans, and, because of their soft deformable bodies, being able to squeeze through or into spaces that humans could not (the US military are already funding two different organisation to develop robots with such deformable bodies).

Like their predecessor in 2025, the robots will be equipped with biometric tools for recognising faces, fingerprints and retinas. It is possible that they will be able to do face recognition with moving targets by this time. They will also be able to conduct on-the-spot DNA testing as well as using heart rate, respiration and temperature sensing to check possible guilt, wrongdoing and lying. They will be able to sniff out explosives and detect concealed weapons such as knives and guns at a distance. They will be our traffic wardens and will police the roads in automated driverless cars. They will be able to sniff out drugs at a distance and condct drink driving tests and they may be permitted to arrest incapacitated individual and return them to the police station for processing by humans.

Many different types of robots will be in use and work in directly connected unison sharing information and images. Using swarm intelligence techniques, everything from tough tracked robots with non-lethal weapons to snake like robots and even pipe crawling robots will operate together to leave no hiding places for law breakers. Teams of robots for specific purposes will quickly and automatically assemble for instant deployment and squidgy robots will be able squeeze into places that human cannot.

Robots will have access to integrated databases of all information on citizens and their friend and associate circles including bank accounts, tax, motoring, shopping, criminal records and movements These will be similar to but much more advanced than the US Total Information Awareness (TIA) and MATRIX (Multistate Anti-Terrorism Information Exchange) programmes that were cancelled due to public outrage. A suspect's movements for particular dates will be checked on the network of surveillance footage.

There will also be many tiny polymer robots equipped with cameras that can be sprayed into populated areas or private dwellings and that can then self-configure into larger robots to constrain criminal activity.

6

Robot Systems

MOBILE ROBOT SYSTEMS

Designing a mobile robot language requires knowledge of the robot's physical configuration and the desired complexity of the implementation. We will define the term "robot" for the purposes of this discussion.

From this definition requirements for a generalized control scheme will emerge. Finally, specific details of the LLAMA system pertaining to robot control will be explained. We have established a message passing requirement-control flows from high to low level while reports flow in the opposite direction. Such a distinction has some impact on the design of a language.

DEFINITION OF AUTONOMOUS MOBILE ROBOTS

A robot is any device which can be physically isolated that can cause a change in its environment in response to sensory stimuli. From this definition, we see that the following can be considered robots: manipulator with force-feedback sensors; miniature maze-traversing autonomous robots. The physical isolation criterion requires that the hardware of the robot can be moved (practicality aside) without upsetting the functionality of the robot.

An arbitrary complex system may have many of the features of a robot and not be a robot: A desk-top computer is not a robot whereas a computer that controls a mobile platform is a robot. The desk-top computer can be referred to as a static tool, whereas a robot could be classified as an active tool.

The presumed robot must have the capability to move through its environment before it can be considered a robot. An automobile cannot be considered a robot since it is a static tool; if the automobile has anti-lock brakes, which sense and affect the environment, albeit slightly, it could be considered a robot. In our society it is difficult to view an automobile as a robot-it really is a mass of metal and plastic hurtling a human occupant to some destination. An automated guide vehicle, like the Sea-Tac automated subway system can be considered a robot. Note that if a mobile platform is controlled by a computer physically removed from the platform, then the platform can be considered a robot and part of a robot system. This is an important observation since this very nearly describes the LLAMA and G2 mobile robot system.

Mobile Robots are a class of robots that have the capability to transport themselves in three dimensional space. As was just observed, the controller need not be physically present on the mobile platform. To describe this difference in configuration, we will define Open Autonomous System as a mobile robot that is controlled by an external system, but not by a person. We will define a Closed Autonomous System as a mobile robot that is entirely self contained.

Using this nomenclature and extending it appropriately, we can describe the LLAMA and G2 project as an Open Near-Autonomous System with capabilities for Closed Autonomy. This says that LLAMA could be programmed to perform some behavior with no external control, although under most circumstances would be controlled by the G2 expert system. The ultimate goal in robotics is an Autonomous System, either open or closed.

An autonomous agent is an abstraction of an autonomous system. Determining whether a system is truly autonomous requires some formal metrics, which is beyond the scope of this discussion. Many models that represent autonomous (robot) agents separate the robot from the environment. Kaelbling presents one possible model of an autonomous system that has the capability for learning by noting that an "agent can be in many different states of information about the environment, and it must map each of these information states, or situations, to a particular action that it can perform in the world" (Kaelbling 1991, 132).

The system she suggests is reactive, which appears to be an essential quality for an autonomous agent. Implementing reactive behavior is trivial; implementing learning is a very difficult problem and will not be covered in this discussion except to note that the LLAMA and G2 systems contain resources that might be configured for learning.

The Hardware-Sensors and Effectors

An open mobile robotic system will have computers with software on and off board the robot. On-board refers to physically residing on the mobile platform, while off-board refers to a location that is independent of the platform. In general, the on-board system will interface with the robot platform hardware and the off-board system will interface with the on-board system. One exception to this generalization is the stationary video camera that connects to the off-board system to monitor and affect the platform's movements. For our purposes, only the hardware that can be accessed by the on-board system is of interest.

A mobile robot system that can analyze and react to its environment through the robot platform is called robo-centric. A robot interacts with the environment by changing its physical presence over time. The robot may change the environment when it enables an effector or senses the environment through its sensors. Its perception of the environment is constantly changing if the robot system senses in a robo- centric manner. For example, even though the robot may not move a great distance and may not change its environment significantly, it is very likely that all sensor readings will have changed their values.

This large-scale sensor update can be a great burden on sensor processing and interpreting systems. G2 attempts to dampen this potential speed bottle-neck by introducing an environmental modeler as part of its knowledge base structure.

In essence the LLAMA and G2 system is robo-centric, but contains an environmental modeler to reduce the complexity of sensor processing. This modeler expects certain environmental qualities as being absolute, such as level floors, ever-present (read: detectable) walls, landmark recognition, etc. These environmental restrictions and modeler are warranted: A person who has memorized how to walk to the bathroom in the dark relying only on feel uses a memorized model of the environment to reach his or her goal. A mobile robot system is much like this person in that the limited sensing capabilities are represented as sensor processing deficiencies.

A direct effector (commonly referred to simply as an effector) in a mobile robot can be a movable appendage (manipulator) or a locomotive device (wheels). Certain sensors can be considered active since they emanate waves or particles into the environment as part of the sensing process. An example of such a sensor is the ultrasonic range finder. The ultrasonic range finder sends a high frequency impulse into the surrounding environment to stimulate a reflection and eventual detection of the reflected wave.

An indirect effector does not change the platform's physical configuration but does affect the environment. For example, an infra-red elevator interface would include a transmitter on the platform and a detector on the elevator control panel.

The robot could request elevator service without making contact with the up/down buttons. If a building containing a robot was "wired" to the robot's supervisor computer, then the supervisor could pass a request to the elevator system directly when the robot and elevator come into close proximity. (Note that the building system-supervisor computer-mobile platform does not conform to our definition of a mobile robot system since the building's elements cannot be isolated without impairing the intent of the design.) The effector which would have been used to control the elevator is no longer needed. The presence of the robot near the elevator and the intent to ride on the elevator will have caused a virtual effector event in the latter variation. There are currently no such events implemented in the LLAMA and G2 system since it is robo-centric.

As robots become more complex the need for self-diagnosis arises. There is no fundamental difference between the robot and its environment to the robot hardware/software and its (robot) environment. A robot is composed of different hardware and software modules that attempt to function as a seamless whole.

If a module fails or falters, it is imperative that the supervisor is made aware of the fact. Each module may contain sensors that determine the condition of the module. Such sensors are no different from the sensors used to sense the robot's environment. We can refer to these as system sensors. These sensors can be further sub-classed, but current robot technology does not include these variations so the one term will suffice. Similarly, system effectors are plausible devices, but are rarely found in a robot. A consistent definition for a system effector is any device which causes a change in another device using a means other than a physically guided interface. An I/O port or optical fiber interface would not qualify under this definition. If these were included in the definition, virtually any software to hardware interface would qualify as an effector! Changes to system parameters or behavior are part of a class called system interface control.

Future mobile robots may contain modules, perhaps miniature repair robots, that monitor and adjust system performance. LLAMA can be configured to produce either solicited or unsolicited system sensor reports. Additionally, a rudimentary "self-repair" can be implemented: if some event causes an emergency stop condition, such as running into an object, a macro can be designed which will determine the stopped condition and reset the emergency-stop automatically.

Commands and Reports

A mobile robot, regardless of any constraints placed on its operation, must have the capabilities for movement and sensing. According to our definition, a robot controlled indirectly by a human would probably report sensor data; however, if the platform contained no sensors, the person controlling the robot would have to substitute his or her sight for this purpose. Another sensor technique that relies on neither the platform sensors nor human sight is known by the term, "God's Eye."

Modern airborne fighter craft display a top-down view of the combat theater for the pilot. Usually this visual information is provided by support radar aircraft and received and displayed in the fighter. This example illustrates an important issue in robotics: Should a robotics system rely only on the platform's sensors or all available sensors in the robot's environment? A robot that has access to global sensors is termed enviro- centric. A properly designed robo-centric robot has the advantage of being completely self-contained and autonomous. If a general mobile robot is required, it could be designed in an enviro-centric manner. A robo-centric system would be autonomous but could enhance its knowledge of the world by obtaining enviro-centric data upon request. Note that the robo-centric system does not rely on enviro-centric data and can function well without it. Since a robo-centric system can only rely on its own sensors (except as noted above), in an open autonomous system sensor data must return from the implementor to the supervisor. This is the basis for reports. Similarly, supervisor to implementor control is achieved through the use of commands. One advantageous side effect of a robo-centric open autonomous system is that the platform can be modeled as a black box. The platform has one input and one output to the stationary system, which lends itself to such an interpretation. A greater difficulty is how to model the environment. Either the environment must be modeled as part of the black box, or the environmental model must somehow interface to the model of the robot's sensors and actuators.

The Mobile Robot System Classes

Most robot hardware can be controlled through software. The hardware classes of effectors and sensors are usually controlled through some software function, either a macro or primitive in the LLAMA system, and executed when the corresponding command arrives from G2. Most of these software functions can be grouped into like classes.

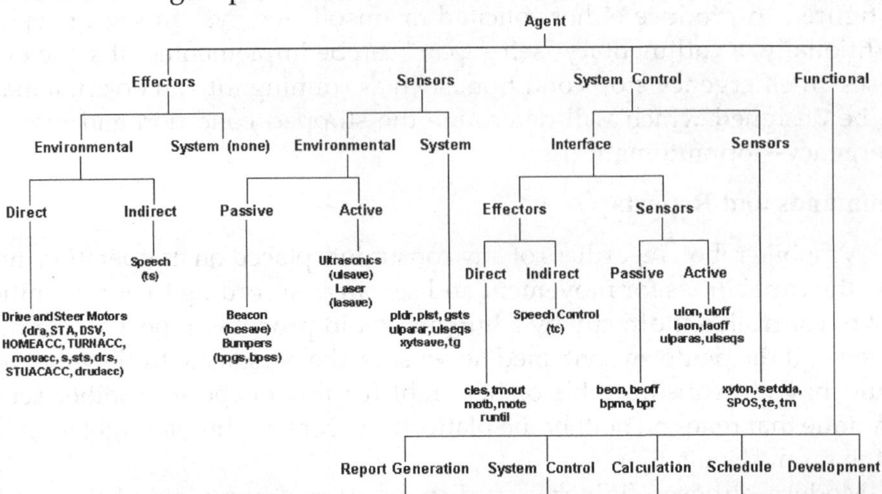

Fig.: Agent Classes. The mobile robot agent classes are presented in a heirarchical format. Due to space limitations, the Functional class has been placed near the bottom of the diagram.

A hierarchical class description is a reasonably clear method of describing the command and report structure of LLAMA; however, dependencies occur across classes and certain of these relationships might be obfuscated. For example, all drive motor commands can be grouped together but are not since each is grouped under appropriate effector, sensor and system headings (classes). An additional class description, called the functional class, supports the programming and operational features of LLAMA. Figure shows the complete class hierarchy. This class description will be expanded upon in the next chapter. The relationship between hardware control through software functions will be covered in more detail in the next chapter.

A MODULAR ROBOTIC SYSTEM AND MODEL BASED CONTROL ARCHITECTURE

Several interesting modular robotic systems have been fielded in recent years, from Legobots and Robix to Polypod and the GMD-Snake. Some designs maximize economy and simplicity at the expense of torque and precision capabilities, while others explore innovative concepts in articulation and control. In contrast, the MMS design seeks to provide contemporary industrial standards of torque and precision in conventional modes of articulation at competitive cost.

In summary, MMS is a general purpose user configurable manipulator system that integrates a modular hardware construction with a hierarchical model-based software architecture for control and programming. The hardware design emphasizes a simple and uniform mechanical and electrical interface among components, competitive performance parameters and flexibility of configuration. The software architecture provides an intuitive graphical user interface for easy manipulator configuration, application programming and simulation, informative task monitoring and multi-subsystem coordination in the workspace.

In the following discussion, a brief overview of application development is presented from the user's perspective, expressing the primary intended mode of system usage. This is followed by a discussion of the Model Manager software, the core of the MMS system. Then, an overview of information flow through the system and a summary of the hardware and software comprising the individual joints. Finally, after a brief review of the command protocol that controls the joints, the discussion concludes with comments about recent improvements to the MMS design and a survey of the advantages offered by this approach.

APPLICATION DEVELOPMENT

Ideally, we may assert that application development, in an industrial or research setting, should consist of a planning and an execution phase. The planning stage would progress from (1) application concept to a (2) workspace and manipulator configuration plan, then (3) task programming and simulation in a virtual setting. The execution phase entails (4) acquisition of the needed components, (5) installation and test of the system and, finally, (6) operation of the system. Funding, presumably outlined in step (1), may be provided incrementally through the first 3 stages, based on assessments made during the process. From step (4) onward, a sufficient funding commitment is critical. By design, the MMS Model Manager software provides broad support for each step of this process, enabling informed decision making and efficient execution of the plan.

In practice, the user envisions a manipulator system, commensurate with MMS module capabilities, that will perform some desired function in a research or industrial environment. Possible target configurations might include special purpose vehicles using rotary joints for propulsion and steering, multi-legged walkers using a combination of rotary and linear joints, bifurcated arms for complex assembly tasks, standard PUMA or SCARA arm geometries performing typical production tasks, simple camera/ antenna gimbals or some combination of the above possibilities.

Then, utilizing the Model Manager's Configuration functions on the system control computer, the user may configure the design concept as a virtual entity for subsequent application programming, exercise and evaluation. Examples of the available module types are shown in Figure, modules are provided with 5, 7, 10, 14 and 20 cm diameters. After evaluating the configuration, the user may choose to assemble the manipulator from real modules, construct the planned workspace environment, install and operate the system.

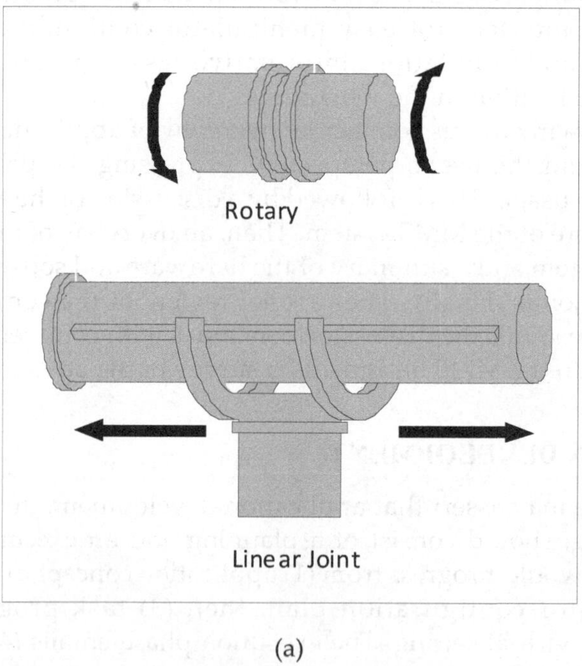

(a)

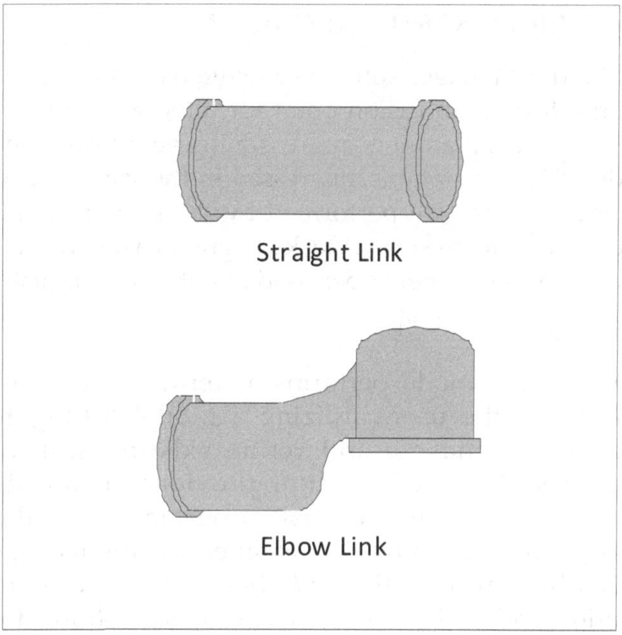

(b)

Fig. a. MMS JOINT module types b. MMS LINK module types.

An example of MMS 7 cm and 10 cm rotary joint modules assembled into a typical 5-DOF arm is shown in Figure. System installation effort may range from minimal (e.g. autonomous vehicle with no immediate site requirements) to very complex and costly installations requiring personnel reassignment, factory layout revision, special purpose tooling and other significant costs.

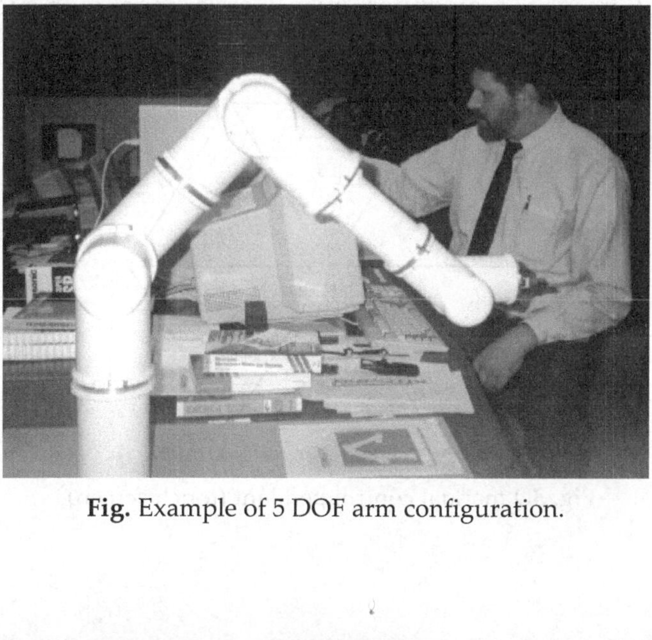

Fig. Example of 5 DOF arm configuration.

MODEL MANAGER CONTROL SOFTWARE

The MMS Model Manager software maintains a central system MODEL that provides an information fusion point for integration of incoming sensor and vision data, adaptive planning and reasoning about problems. The system MODEL includes TASK programs, expressed in the various primitives of the MMS Task Language, that are performed by the system manipulators. The Model Manager can perform several task programs asynchronously and each individual task operates in either supervised or autonomous task control mode.

Task control modes

- Supervised task mode performs a stepwise execution of the task which allows the user, utilizing a Model Editor, to create task definitions or analyze and refine existing task or workspace components. The Model Editor provides an intuitive graphical interface for constructing task programs from the MMS Task Language and also provides an interface for teaching workspace locations by directing the end-effector to a desired location and orientation and assigning it a name. Figure shows the 3D manual control pendant used to teach locations. When rendered within the workspace model there is no need for reference to coordinate frame axis labels, the user simply clicks (or touches) the arrow pointing in the direction that the end-effector must move. Activation at either end of an arrow produces large movements in the corresponding direction, near the middle movements are small and slow for fine adjustment of locations. These positions may then serve as reference points for use with information from a CAD database or high level primitives of the Task Language. The www.mms-robots.com web site offers visitors an opportunity to experiment with such a pendant in a simulated environment.

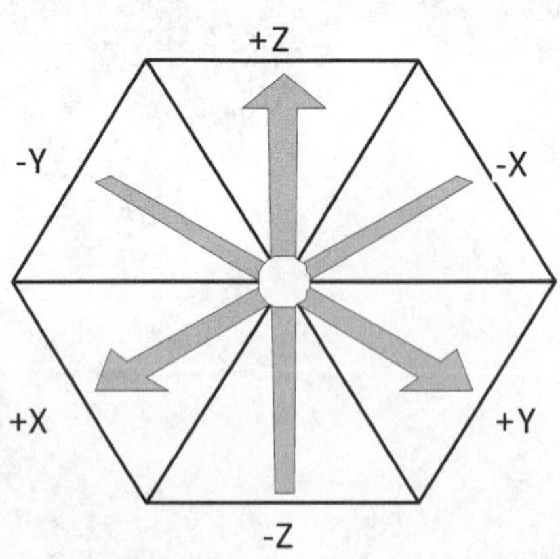

Fig. 3D manual control pendant (touch screen).

- Autonomous task mode performs a TASK without user intervention. The user may, however, invoke the Model Editor to modify various components of the system model dynamically, as desired.

Execution modes

In either control mode, tasks may be performed in virtual or real execution mode.

- In Virtual execution mode, the user selects one or more virtual viewpoints for observation and performance of the task in a simulated workspace is rendered in corresponding display windows on the system monitor for analysis. The task is performed with a snapshot copy of the system model, motion commands are sent only to the simulated manipulators in the model snapshot and sensory inputs are obtained only from model components. Representation of basic physics in the model (collision detection, dynamic behavior, etc.) is an important goal of future development efforts.
- In Real execution mode, commands go to the real manipulators and the sensory inputs update the appropriate parameters of the system model in real time where possible.

OVERVIEW OF MMS OPERATION

An overview of a simple system configured with 1 arm assembly and a desktop arm control processor (ACP). When operating, arm commands originating from the TASK program running in the ACP, traverse the common serial communications bus to reach all joints of all arms concurrently. Each joint receives the same command string, determines its component of the commanded motion, generates an appropriate trajectory interpolation and applies classical control techniques to follow that trajectory. The hardware and software implementation of this functionality is now discussed in greater detail.

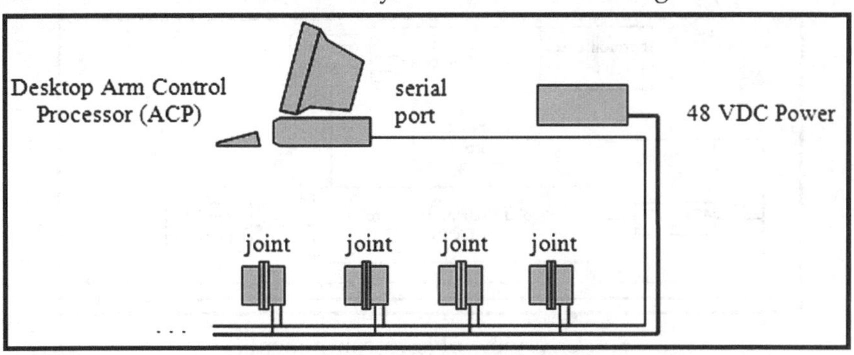

Fig. MMS system overview.

Only two connections are made to each joint: +48VDC power bus and the serial communication line, both connect automatically when modules are assembled. All joints draw as much current from the 48V bus as required to drive their respective servomotors. The ACP serves as a strict communications bus master. Generally, joints are listening for motion commands and polled en mass for problems. Only if there is trouble does the ACP master instruct the individual joints to report their status, allocating a brief time window for each.

JOINT IMPLEMENTATION

All motion in an MMS system emanates from the self-contained one degree-of-freedom JOINT module which, rotary or linear, has the same simple mechanical and electrical interface to the rest of the system as every other MMS component. The joint modules, however, listen to the communication bus passing through and act accordingly. Each joint contains an imbedded 16-bit control processor that receives, interprets and executes commands, a high precision joint position sensor, a DC power amplifier, DC servomotor and harmonic drive speed reduction drive gear. Figure presents an overview of the subsystems comprising every joint. To command all joints concurrently with minimal serial bus traffic and to provide a high level of path and velocity control, a flexible and concise high-level joint module control protocol (JCL) has been developed. Current joint design achieves competitive cost, force, speed and precision parameters. Joints also provide process I/O ports for connecting the MMS units to the workspace for analog and digital signal sensing or process control. A few aspects of joint module implementation are discussed below.

Joint Software Organization

The functional organization of the joint is reflected by the structure of the program running in the joint control processor (JCP), one of which is present in every joint module.

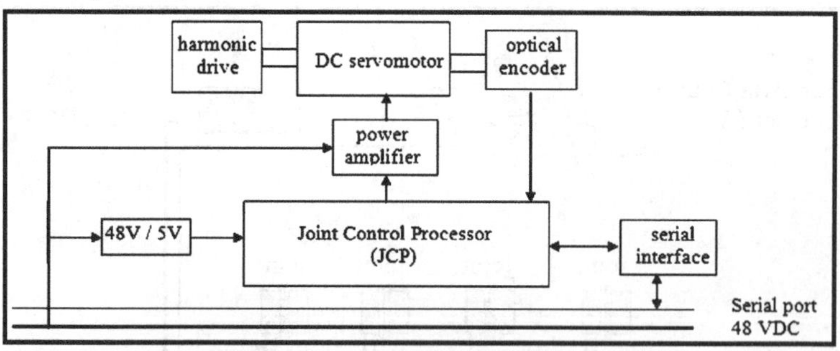

Fig. Overview of joint components.

The processing activity occurs on 3 interrupt service levels:

- *Level 1:* On the lowest interrupt priority level (background), ASCII command strings are processed from the incoming message buffer, parsed and error checked. If good, the commands therein are executed, see the discussion of the JCL below. For motion commands this implies that the arm solution function is computed for this joint by the FAM (fuzzy associative memory) subsystem developed for MMS to accommodate variable arm geometry, see the FAM discussion in section 7.4. For each Cartesian pose vector, pose vector time derivative and associated time interval (13 elements) broadcast to all joints, a unique 3-element vector (joint angle, angular velocity, time) is placed on each joint's position queue.

- *Level 2:* ASCII character command strings arriving over the communications bus are assembled in the incoming message buffer. Input is screened for commands that are executed immediately (e.g. Emergency Stop, etc.).
- *Level 3:* At the highest interrupt level, a hardware timer cycling at 500 - 1000 Hz performs the calculation described in Figure, consisting of two parts:
  - The top two entries in the position queue define a cubic polynomial interpolating function, see the discussion of Hermite Polynomials in reference 5. The real time clock measures the time into the interval and from the cubic, a function value (angle) and a derivative (angular velocity) are calculated for use in part (b) as the schedule model. When not too near to singularities, this method provides good path and velocity control (i.e. circular arcs, straight lines) and continuity from relatively few control points.
  - The joint position, as defined by the optical position encoder, is recorded at each interrupt. The last two values are used to estimate the position and velocity of the joint, referenced as the sensor model. These two state vectors are differenced to produce the error model. A classical PID control algorithm then calculates the torque required from the servomotor to optimize compliance with the planned trajectory. This, in conjunction with the estimated back EMF from the servomotor (motor model), determines the output voltage parameter placed in the motor drive PWM (pulse-width-modulation) register.

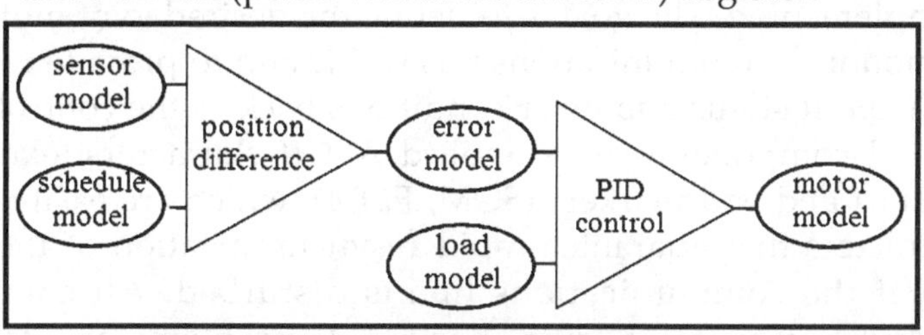

Fig. Trajectory control algorithm overview.

Joint Hardware

Several of the subsystems described in Figure merit further clarification. The joint control processor was implemented with an Intel 80196KR, which provides several convenient features that kept the on-board chip count to a minimum:

- The quadrature type optical encoder attached to the servomotor shaft connects directly to the processor which provides decoding logic that drives an up/down counter at several MHz, with no processor overhead.
- The built-in serial port interface capable of data rates well above 100K baud.
- The built-in pulse-width-modulation (PWM) port (plus one additional polarity line) provides the type of signal needed to drive the power amplifier.
- The 16-bit architecture provides the needed computing power for calculating the solution functions, generating the position interpolation cubic and calculating the PID control loop in real time at 1KHz.
- Sufficient on-board RAM for all parameters, registers, position queues and message buffers.
- Sufficient on-board ROM for several thousand lines of C code (cross-compiled with Archimedes 80196 development software).

The IR8200 power bridge driver permits delivery of 300W bipolar continuously variable voltage (-48VDC to +48VDC) to the servomotor without generating significant waste heat and using only a few cm3 of space. For higher power requirements, multiple boards may operate in parallel.

The harmonic drive speed reduction output gear is small, light, efficient and most importantly, has minimal hysteresis, which makes possible the high resolution measurement of joint position with a simple optical encoder on the motor shaft. Unfortunately, they are expensive, representing about half the production cost of the joint.

JOINT CONTROL LANGUAGE

To simplify and streamline the joint command process a control language has been implemented. The goal is to define the desired joint motions with a minimal amount of communications bus traffic and to provide rapid flow of information about status and error conditions back to the control computer. All command components are standard ASCII characters, except for the command start and end markers (SOM, EOM), which are easily recognized special characters that guarantee rapid resynchronization of the command interpreter if the communications line is disturbed. All commands are vertically and horizontally error-checked. If problems are detected, the command is discarded and the error is reported to the ACP at the next status poll. The general command format is:

SOM	Start-of-message control character
CHK	Checksum for entire command
ARM	Command target arm (*=all)
JNT	Command target joint (*=all)
CMD	Command
PARM*	Command parameters

Some of the most frequently used commands are:

Command	Code	Function
RESET	I	Reset/ reinitialize all joint parameters (used at startup).
MOVE		M Move the end-effector through the specified sequence of positions.
SET	P	Set operating parameter. First argument specifies parameter (PID coefficients, command modes, etc.), the second specifies the value.
STATUS	S	Report joint status.
HISTORY	H	Report joint error history.

Positions are defined as 13-element vectors. The first 6 parameters define a position (x, y, z, ?x, ?y, ?z), the next six define a rate of change (linear and angular velocities) and the last specifies a time (measured from the previous position in the sequence) at which the end-effector is to arrive at that location.

RECENT MMS IMPROVEMENTS

Experience with the prior generation of MMS units has brought to light several opportunities to increase performance and reliability while decreasing cost and complexity.

Connector Latch

The mechanical interconnect on prior MMS units employed a circumferencial clamp and index pin design. If the mating surfaces were not clean or the clamp not properly tightened there could be slippage and positional accuracy would be compromised.

To eliminate this frailty, a simple auto-lock interconnect latch has been developed that provides a strong and rigid linkage and shows good immunity to improper assembly.

Brushless Joint Motor

The use of brushless motors in the joints will provide higher power density, lower reaction torque and longer service life without maintenance. The MC model of the Intel 80196 processor provides operating logic and control registers for brushless motors. A prototype unit is in development for evaluation.

Single Control/Power Line

Prior implementations have employed 2 conductors (plus ground) for the system bus: power and communication. Since each joint's driver amplifier must be RF isolated from the power bus anyway, it is convenient to superimpose system communications on the DC power bus on the power source side of the RF isolators.

This is advantageous because the number of slip rings in each joint is reduced by half and the module electrical interface needs half as many terminals. The disadvantage is that all joints need RF detectors and drivers. This is also part of the new prototype design.

FAM Solution Function

The MMS architecture presents unique challenges to the usual arm solution function determination process. The user may configure a geometry for which no solution has ever been developed, or worse yet, for which no solution is possible for certain end-effector positions. Early MMS systems addressed this problem by configuring standard PUMA and SCARA designs.

To eliminate these restrictions, a Fuzzy Associative Memory (FAM) solution function generator has been developed. Tests have shown adequate speed and precision at runtime and the user is not encumbered, see references 7-10. The FAM system has a setup phase and a runtime phase:

At setup: The user has defined the arm configuration for the system model and the FAM system uses the resulting arm model to construct a fuzzy rule set that provides the desired accuracy over a specified region of the workspace. This procedure may run for some time on the desktop computer so the user is advised to limit the requirements accordingly.

At runtime:The fuzzy rule set is used at run-time to generate a solution for each control point dynamically. It is preferable to distribute a scalar version of this computation to the joints, relieving the desktop host of this burden, but this requires inclusion of a flash RAM chip in the joint hardware. Continuing escalation of desktop computing speeds has made this unnecessary.

Teleoperation Extension

The joint command set has been expanded to include support for teleoperation modes of operation. These commands differ from the normal mode of control (using long sequences of predefined positions) by rapidly updating a single position-velocity 12 element motion state vector to optimize compliance with operator input. In the absence of sustained input, the joint linearly reduces the velocity component of the motion vector to 0 magnitude within a specified time constant.

SIMULATING ROBOT SYSTEMS

How can a macro or composition of performance orders be verified for proper operation in LLAMA? Performance orders will eventually be generated by G2 with no user interaction.

This chapter presents (1) a simplified finite state machine representation of LLAMA, (2) a "walk-through" of a simple macro for operational verification using a proposed consistency checker, (3) a suggested graphical method of representing LLAMA commands in G2, and (4) the results of two performance orders executing in parallel. Item (4) could be used to verify any future consistency checkers.

REPRESENTING LLAMA WITH FINITE STATE MACHINES

Portions of LLAMA can be represented using Finite State Machines (FSM). FSMs can be used for simulation purposes. Achieving a thorough representation of concurrent LLAMA operation would require a technique such as Petri Nets, which is beyond the scope of this discussion (Murata 1989, 541-580). FSMs can be accessed when system states are needed in a LLAMA command simulation: Since many LLAMA commands activate and deactivate hardware and software modules in a binary fashion, a two state FSM is appropriate for simulating this behavior. Trajectory simulation can be performed using a variation of the algorithm presented in appendix A. Sensor simulation can be performed using similar techniques. The following states illustrate how the drive and steer motion status of the robot can be represented using FSMs:

- Dr0 = Drive motor stopped (initial)
- Dr1 = Drive motor turning
- St0 = Steer motor stopped (initial)
- St1 = Steer motor turning
- ES0 = System motion ready (initial)
- ES1 = System in emergency stop
- ME0 = Motion Enabled (initial)
- ME1 = Motion Disabled

A FSM diagram of robot drive and steer motions is given in figure. Note that if the system enters state ES1, this implies that a transition to states St0 and Dr0 occurs. If the system enters state ME1, all Dr and St state transitions are disabled. In a simple example using two performance orders, we can analyze the behavior of the system based on the above states. Assume the following performance orders have been transmitted:

- P.O. #1: { 10 30 00 3 txabs 20 st ststop }
- P.O. #2: { 10 30 01 3 txabs plst 0 = if sts endif }

Before analyzing the behavior of these performance orders, the components need to be explained:

- 10 30 00 3 txabs-waits until 10:30:00 am arrives, then executes next commands.
- 20 st-turns head clockwise at 20 degrees per second.
- ststop-waits until head stops turning.
- plst-returns 1 if the head motor has stopped, 0 if it is still turning.
- 0 = if sts endif-a head motion stop is issued if plst returned a zero.

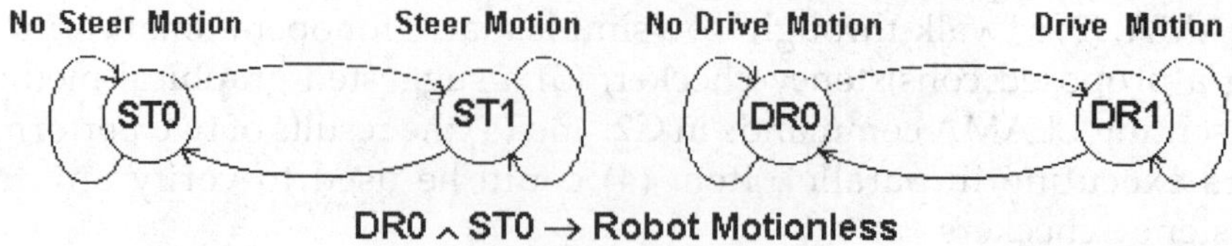

Fig. Finite State Machine Description of Motion. Drive and steer motions on the Denning Mobile Robot are independent.

The question is: will both performance orders complete or will one or both be temporally bound forever? The presence of "ststop" indicates that the upper performance order could conceivably wait forever for the head motion to stop since it contains no means of stopping the head. The first performance order will start at 10:30:00 AM and the second at 10:30:01 AM. We note that initially there is no motion (Dr0 Ù St0) in the system and that all motion is enabled (ES0 Ù ME0). The following events take place:

10:30:00	Steer motor slews head at 20 degrees per second State change: (St0 -> St1)
10:30:00+	Performance Order #1 waits until head motion ceases (Loop until St0)
10:30:01	Performance Order #2 checks if in state St1, if yes, St0 and end State Change: (St1 -> St0)
10:30:01+	Performance Order #1 ends since state St0 reached

Both performance orders complete at about the same time-one second after 10:30 AM. If an emergency stop occurred during execution of the first performance order, the first order would become temporally bound forever. Since performance orders announce when they start and stop execution, G2 could monitor the status of the performance orders and issue an "exit" instruction to the first performance order if it has exceeded its implied execution time of one second. A LLAMA simulator in G2 can use FSM's to estimate the progress of performance orders and take corrective measures if it becomes necessary.

CREATING PERFORMANCE ORDERS: THE CONSISTENCY CHECKER

Once a LLAMA performance order has been created, it can be analyzed using a proposed G2 module that analyzes the program structure for consistency. This module will verify the following qualities of the performance order: (1) commands must have a stack consistency, implying that parameters are not lost or added unnecessarily to the stack; (2) certain instructions that are not temporally bound must not occur in sequence and certain instructions must occur in groups-the consistency checker can verify, warn, or fix most of these cases; (3) ensure that the program will do what it is expected to do under all or most circumstances. The following example illustrates some of the issues in designing a consistency checker for LLAMA performance orders.

The following scenario is presented: G2 needs to generate a performance order with a performance that will move the robot in a 180 degree arc clockwise from the robot's current position and orientation with some unspecified radius. G2 performs a search which locates no performances with the required qualities. G2 proceeds to create a performance from various skills using a performance generator. Unfortunately, the hypothetical performance generator occasionally makes errors in its creation of performances. The following is the performance created from this generator. Note that the indentation and lines are for readability but the performance is a linear sequence of operators when sent to LLAMA as a performance order.

```
0 0 0 spos                              (set encoders to zero degrees)
3 5 dra 200 100 STA                     (start motors at arbitrary speeds and accelerations)
begin
                    xytsave             (get position)
                    3 rem               (remove unneeded data)
                    0 >                 (test for completion)
until
s                                       (stop motors)
```

Let's analyze the performance. State variables and stack pointer need to be initialized:

Steer = St0
Drive = Dr0
E-Stop = ES0
Motion = ME0
stack index i = 0;

The proposed algorithm to validate the performance is as follows:

Stage 1: Validate stack requirements:

```
(push(i=1) 0; push(i=2) 0; push(i=3) 0; x pop(i=2);y pop(i=1) )
(t pop(i=0); spos; push(i=1) 3; push(i=2) 5; ds pop(i=1) )
(da pop(i=0); dra; push(i=1) 200; push(i=2) 100; ss pop(i=1) )
(sa pop(i=0); STA; begin; xytsave; push(i=1) th; push(i=2) y)
(push(i=3) x; push(i=4) 3; push(i=5) 3; num pop(i=4); rem)
(i=i-3=1 (num); push(i=2) 0; c1 pop(i=1); c2 pop(i=0);>)
(push(i=1) 1/0; test pop(i=0); until; s)
```

We note that i equals zero after this sequence which means that no numbers were left on the stack. Also, the stack index never dropped below zero, indicating that an attempt to remove a non-existent number from the stack never occurred. Since "xytsave" may leave either one or four numbers on the stack the previous algorithm needs to include branches for all possible stack items that an operator may return. This type of extended testing could use a search tree that would work its way through the arguments and whenever an operator may produce more than one set of values, each branch in the tree could be tested. Note that in this tree the branches fall back into the roots implying a potentially very efficient algorithm for solving this problem.

Stage 2: Validate temporal binding and groupings

(spos - temporally bound)

(dra - a temporally unbound command, valid, no other dra before)

(STA - a temporally unbound command, valid, no other STA before)

(begin - valid, first begin in sequence)

(xytsave - note: not grouped, no xyton at start of sequence)

At this point the consistency checker will automatically insert an "xyton" at the beginning of the performance.

(rem, >, ignore)

(until, valid, closes begin loop grouping)

(s, valid, resets motion state machine to original state)

(a flag alerts the system that an xytoff is needed at the end to match inserted xyton)

The system is valid in the second stage of the tests.

The final, and most difficult part of the algorithm is deciding whether the begin-until loop will be exited under all conditions. As noted in the previous section, if the motors are engaged and an emergency stop is issued, loops must be verified so they won't hang (although this condition cannot always be guaranteed).

Stage 3:

The consistency checker determines that the performance has no means of detecting an emergency-stop condition, implying the performance might enter a suspended state. To remedy this situation, the consistency checker inserts a function that puts a "1" on the stack if the system is stalled due to an emergency stop or it places a "0" otherwise. It also adds an "or" function so that either the check for 180 degree rotation is tested or stalled motion is tested. The final, corrected algorithm is as follows:

```
xyton                       (initialize encoder read)
0 0 0 spos                  (set encoders to zero degrees)
3 5 dra 200 100 STA  (engage motors)
begin
        xytsave                     (get position)
        3 rem               (remove unneeded data)
```

```
        0 >                          (test for completion)
        pldr                         (test for stall)
        or                           (combine results)
    until
    s                                (stop motors)
    xytoff                     (disable encoder read)
```

Unfortunately, this algorithm will not work because the "greater-than" test is incorrect, it should be "less-than." The performance generator, which was not well developed, used the wrong symbol in the loop. The function "xytsave" puts an angle between -180 and 180 degrees on the stack. If the steer motion is positive, the angle returned will be positive until the robot has turned 180 degrees, at which point the returned angle will suddenly switch to -180 degrees and increase until it reaches zero.

If the steer motion is positive, the head will have turned 180 degrees the moment the returned angle switches from +180 to -180 degrees. Thus we need only wait until the returned angle is -180 degrees or negative to complete the 180 degree turn. The "greater-than-zero" test should be "less-than-zero" if we want a negative angle. As listed above, the loop will be exited immediately since the returned angle will be positive during the first 180 degrees of motion.

This problem could be detected by sending the performance order to a simulator. If the simulator produces an output which is not what is expected, the consistency checker could request that a user fix the problem in the performance generator.

Problems could be isolated by observing the simulator's real-time trajectory output. Ideally, the expert system could also be used to store a repertoire of skills with known behaviors. When these skills are built into a performance, the final product could be verified using a consistency checker.

VISUAL PROGRAMMING FOR LLAMA

LLAMA performance orders are serial entities: A performance order is received one character at a time. A graphical composition system could be used to define how such a string is created. G2 has the capability to express LLAMA commands in a graphical format. If these commands could be expressed as connected objects, each object could contain attributes that define how it operates in the LLAMA framework. Such an approach could become part of a simulator and consistency checker. This visual programming interface, along with a simulator and consistency checker could aid in the understanding of how LLAMA operates and would produce elaborate code without the burden of using a text editor.

We present a method in which LLAMA can be encoded into G2 objects in this section. If these objects include certain attributes, a visual programming system can be developed, albeit one in which a user must know FORTH and LLAMA fundamentals. Appendix B provides the attribute descriptions for all commands in the LLAMA system.

Fundamentals

A performance order is a set of instructions sent to the robot. In G2 a performance order object could contain attributes that determine when the performance order is to execute, and when executing, its current execution status.

This is very important since it identifies that LLAMA and G2 are operating in a symbiotic manner. For example, we can represent whether the robot is moving, by incorporating a logical variable in G2 that represents the two state FSM described earlier in this chapter. The performance order execution status attribute in G2 could be used to verify completion of the order or it might confirm that the order has stalled or locked-up. The execution attribute receives notice from LLAMA when the performance order has started so that G2 can know when to send another performance order.

The performances in a performance order, since they are a linear list of commands, can be reduced to macros. These macros could be defined using a definition order: a semi-permanent list in the on-board system of commands represented by a single token.

The implication is that once a definition order creates a macro, which might take a minute to download, then it is available for immediate use by a performance order. If it is known that a performance order will be executed at some time and the system is idle at that instant, then the performance can be sent as a definition order. Figure illustrates the advantage of a definition order when transmission times are restricted due to the limited baud rate of the radio link.

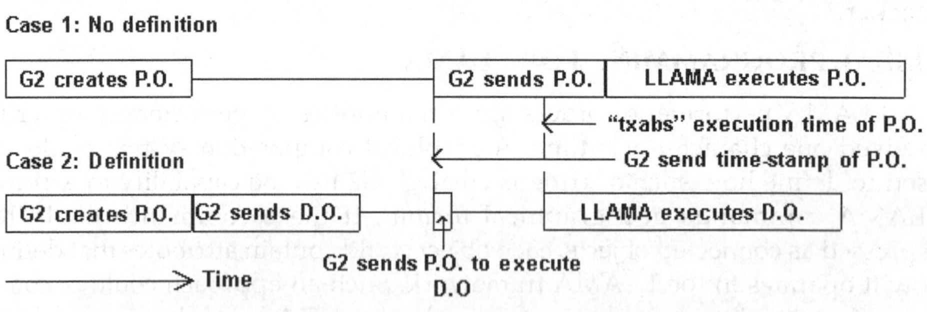

Fig. Performance and Definition Order Execution Sequence. If G2 makes the false assumption that transmission time of a performance order is invariant, then if transmission time is increased, the performance order will very likely execute late. This is shown in case 1. If a performance order is sent as a definition order, there is more flexibility in the transmission time, since the bulk of the performance was sent earlier, as shown in case 2.

Once a performance order string has been received, LLAMA creates a spawned process that will execute the instructions. LLAMA returns a handle number which uniquely identifies the performance order and G2 places that number and a time stamp into the performance order object's execution status attribute. After receiving the performance order, LLAMA attempts to compile the performance order string. If an error occurs during compilation, G2 is notified and a no-handle message is returned, indicating that the performance is not executing and the compilation failed. If all was successful the performance order will start executing. Once execution is complete, a termination message and the performance order's handle number is returned to G2. It is essential that G2 keep track of which performance orders are active, otherwise G2 may conclude incorrectly that a performance order has completed when it may not have.

The LLAMA Primitive Object

We have seen that certain elements of LLAMA can be simulated in G2 using a FSM. Sensors and actuators could be simulated using more elaborate techniques. The simulators and FSM could be interfaced into the LLAMA primitive commands using an object representation. A command can be described in G2 through a hierarchy of G2 objects. Figure shows a typical primitive, "movacc," with its operational qualities exposed. The upper most object, the "execution object," represents an instance of "movacc," with connection stubs to connect it to other execution objects (primitives and macros). The execution object contains the "movacc" parameters, in the form of "number" objects (push object), and the "movacc" object itself. The "number" objects precede the "movacc" object because every primitive object representation in G2 will mirror the linear flow of commands to LLAMA. The execution objects are designed such that, given the appropriate G2 rule and procedure controls, a connected sequence of these G2 macro and primitive representations can be analyzed for consistency or incorporated into a simulator.

In this connected object scheme, a "daemon" could be used to move along such a connected system and would "activate" and operate upon every appropriate object. If the selected object is "number," then the value of the object is placed ("push") onto the stack (a subclass of the list object). When the "movacc" object is activated, control descends to an object lower in the execution object's hierarchy that contains "un-number," or "pop" objects. These objects are connected to the "movacc" interface control object. When each is activated, it pops a number from the stack and inserts it into the given attribute of the interface control object. Then once the "movacc" interface control object is activated, it calls a procedure that analyses the simulated LLAMA system's state and modifies it appropriately. Control continues to the next execution object.

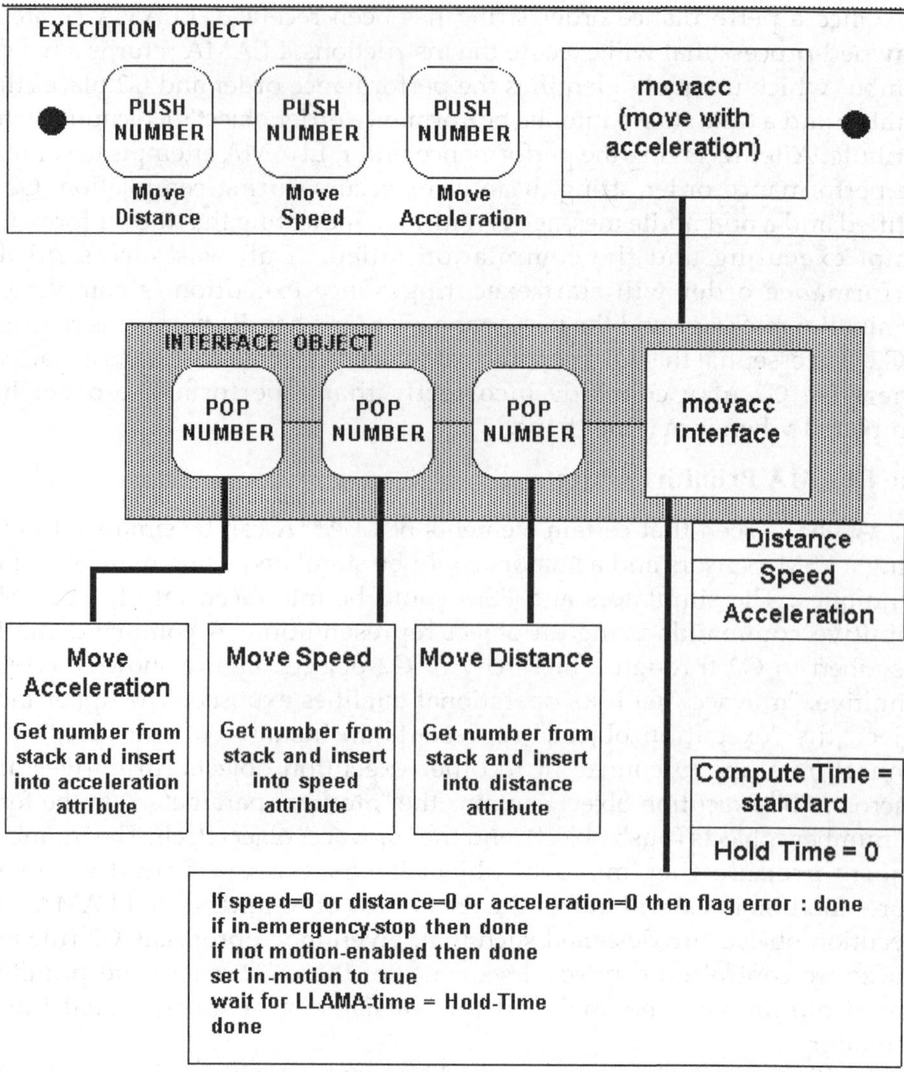

Fig. An Execution Object Representation of "movacc." A hierarchy of objects is used to form a complete specification for the "movacc" command.

It may seem that the push and pop objects could be represented as one, simplifying the entire object representation; however, under certain circumstances instructions, such as "movacc," might obtain parameters from the stack placed there by other primitives or macros. The pop numbers are not alterable since they control the particular primitive's interface to the simulator whereas the push numbers could be removed and the stack values implicitly obtained from previous execution objects. The following is an example of such a case:

```
10 1 do I 5 5 movacc drstop loop
```

Although "movacc" requires three parameters, the first parameter is obtained implicitly from the stack as placed there by the "I" operator, which obtains the do loop's current index and places it onto the stack. In such an instance, the "movacc" function would no longer have three push numbers, but only two, in its execution object.

VISUAL PROGRAMMING

In standard structured programming, a programmer will enter command lines in a text format, typically employing indentation to provide improved code readability. Indentation or some form of delimiting is generally used to set a control structure apart from code.

In the proposed visual system, a different control structure can be developed if macro and primitive execution objects are placed on individual workspaces and connected together to form skills and performances. By using G2's capability to show the workspace hierarchy, an object representation of a program can be produced.

All LLAMA commands can be composed under a visual programming system. Most of the commands are linear and lend themselves to an object representation. Conditionals and variable definitions require additional support, but still can be made to fit the visual framework. Figure provides a basis for this design. Appendix B lists the attributes of each LLAMA command to help translate them into a visual style. A detailed explanation of visual programming is presented in Chang (Chang 1990).

CONCURRENT PERFORMANCE ORDERS: INDIRECT INTERACTION, AN EXAMPLE

We wanted to show that it was possible to create complex system behavior by using only LLAMA commands. The chosen example is a "beacon follower." A person holds an infra-red beacon in front of the robot. The robot attempts to approach the beacon but is allowed no closer than two feet. The robot must also rotate to follow any left-right motion of the moving beacon. A definition order approach will be used.

Performance order number 1 turns the head of the robot so that it always faces the beacon. Performance order number 2 moves the robot forward and uses the ultrasonic range finder to ensure that the robot stops if it gets nearer than two feet from the beacon and the person holding it. (Note that the beacon could be mounted on the other robot.) A pseudo-code version of the performance orders followed by the LLAMA code is given below (refer to the LLAMA User's Manual for a detailed description of the commands):

P.O. #1: Turn Head to Track Beacon

1. Loop until a beacon is detected. 2. Get azimuth angle to beacon (+/-10 degrees). 3. If the distance is < 0.2 degrees, stop steer motor. 4. If distance is >= 0.2 degrees, slew motor in proper direction to center beacon.

Implementation of P.O. #1:

```
: bfollow
beon begin                        (turn on beacon blackboard and begin loop)
    besave                        (get beacon data)
    if                            (execute if good data)
        1 > if                    (execute if found beacon)
            drop rot drop swap drop        (get angle)
            bturn                 (execute motion)
        else
            4 rem                 (drop beacon data, since no beacon found)
        endif
    else
        sts                       (stop steer motor due to bad beacon data)
    endif
poll @ until                      (check "quitting" variable, continue until
done)
sts                               (turn off steer motor, just in case)
beoff;                            (turn off beacon blackboard)
: bturn
dup                               (duplicate angle)
abs 2 <                           (0.2 degrees, check if centered)
if                                (centered)
    plst 0 =                      (check if turning)
    if                            (turning)
        sts                       (stop steer motor)
    endif
    drop                          (drop extra angle value)
else
    3/ st                         (use extra angle to turn at speed prop. to
angle)
endif;
```

P.O. #2: Move Forward Unless Less Than 2 Feet

1. Get current head angle
2. Get three ultrasonics pointing in forward direction
3. Get minimum distance of the three sensors
4. If distance is <= 2 feet, stop drive motor
5. If distance is > 2 feet, move robot forward slowly

Implementation of P.O. #2:

```
: mover
ulon xyton                        (turn on ultrasonics and position blackboard)
begin                             (start loop)
    1000 minim !                  (initialize minimum distance variable)
    make3                         (get the three forward ultrasonics numbers)
```

```
        getmin                     (get the minimum distance of the three
ultrasonics)
        20 > if                    (distance is greater than 2 feet)
            pldr if                (check if not moving)
                2 dr               (drive forward if not moving)
            endif
        else
            pldr 0 =               (check if moving)
            if
                drs                (stop drive motor if not moving)
            endif
        endif
        poll @
    until                          (check if time to quit)
    drs                            (stop drive motor)
    uloff xytoff;                  (turn off ultrasonics and position blackboard)
    : make3 GANGP                      (get head angle)
    150/                           (get a number between 0-23)
    dup                            (again)
    1 -                            (is it 0 ?)
    0< if                          (yes)
        drop 23                        (left adjacent ultrasonic number is 23)
    endif
    left !                         (save left)
    dup                            (get altered head angle)
    1 +                            (is it 23 ?)
    dup
    24 = if                        (yes)
        drop 0                     (right adjacent ultrasonic number is 0)
    endif
    right !                        (save right)
    center !;                      (save center)
    : getmin
    ulsave                         (save 24 ultrasonic readings)
    drop                           (assume all readings valid so drop count)
    24 rep 24 rep                  (duplicate readings twice)
    left @ getval                  (is left reading smallest?)
    right @ getval                 (is right reading smallest?)
    center @ getval                    (is center reading smallest?)
    minim @;                       (get calculated minimum distance reading)
    : getval
    -1 *                           (get ultrasonic number 0-23, make into -23 to
0)
```

```
        24 +                    (add 24 to reverse sense)
        deep                    (extract distance reading at depth = ultra
number)
        dup                     (save for later)
        minim @ <               (check if smallest distance)
        if                      (yes)
            minim !                 (save as smallest distance)
            23 rem              (drop all remaining readings)
        else
            24 rem              (not smaller, drop all readings)
        endif;
```

To execute these performance orders, the following sequence is performed:

1. Initialize any variables

```
        1000 variable minim
        0 variable left
        0 variable center
        0 variable right
        0 poll !            (enable loops)
```

2. Send all definition orders (in this order)

```
        : bturn...;
        : bfollow...;
        : getval...;
        : getmin...;
        : make3...;
        : mover...;
```

3. Initialize robot position

```
        home ststop 0 0 0 spos
```

4. Execute performance orders

```
        { bfollow }
        { mover }
```

5. Terminate when done

```
        1 poll !
```

When these performance orders are executed, the robot will wait until it finds a beacon. It will track to it and approach it. Initially, if no beacons are found but are present, a "10 st" steer command could be placed after step 4 to search for beacons; the performance order number 1 will override the steer command once a beacon is found.

Notice that the two performance orders do not share any variables (other than "poll") and rely only on the interaction of the robots sensors and effectors. Each performance order is independent of the other, but a coordinated behavior emerges from their execution.

This shows the "indirect" interaction of performance orders.

7

Space Mobile Robots

ROBOTIC SPACECRAFT

Fig. An artist's interpretation of the MESSENGERspacecraft at Mercury

A robotic spacecraft is a spacecraft with no humans on board, usually under telerobotic control. A robotic spacecraft designed to make scientific research measurements is often called a space probe. Many space missions are more suited to telerobotic rather than crewed operation, due to lower cost and lower risk factors. In addition, some planetary destinations such as Venus or the vicinity of Jupiter are too hostile for human survival, given current technology. Outer planets such asSaturn, Uranus, and Neptune are too distant to reach with current crewed spaceflight technology, so telerobotic probes are the only way to explore them.

Many artificial satellites are robotic spacecraft, as are many landers and rovers.

HISTORY

The first space mission, Sputnik 1, was an artificial satellite put into Earth orbit by the USSR on 4 October 1957. On 3 November 1957, the USSR orbited Sputnik 2, the first to carry a living animal into space – a dog.

Ten other countries have successfully launched orbital missions using their own vehicles: USA (1958), France (1965), Australia (1967), Japan and China (1970), the United Kingdom (1971), India (1980), Israel (1988), Iran (2009), and North Korea (2012).

DESIGN

In spacecraft design, the United States Air Force considers a vehicle to consist of the mission payload and the bus (or platform). The bus provides physical structure, thermal control, electrical power, attitude control and telemetry, tracking and commanding.

JPL divides the "flight system" of a spacecraft into subsystems. These include:

Structure

This is the physical backbone structure. It:

- Provides overall mechanical integrity of the spacecraft
- Ensures spacecraft components are supported and can withstand launch loads

Data handling

This is sometimes referred to as the command and data subsystem. It is often responsible for:

- Command sequence storage
- Maintaining the spacecraft clock
- Collecting and reporting spacecraft telemetry data (e.g. spacecraft health)
- Collecting and reporting mission data (e.g. photographic images)

Attitude determination and control

This system is mainly responsible for the correct spacecraft's orientation in space (attitude) despite external disturbance-gravity gradient effects, magnetic-field torques, solar radiation and aerodynamic drag; in addition it may be required to reposition movable parts, such as antennas and solar arrays.

Telecommunications

Components in the telecommunications subsystem include radio antennas, transmitters and receivers. These may be used to communicate with ground stations on Earth, or with other spacecraft.

Electrical power

The supply of electric power on spacecraft generally come from photovoltaic (solar) cells or from a radioisotope thermoelectric generator. Other components of the subsystem include batteries for storing power and distribution circuitry that connects components to the power sources.

Temperature control and protection from the environment

Spacecraft are often protected from temperature fluctuations with insulation. Some spacecraft use mirrors and sunshades for additional protection from solar heating. They also often need shielding from micrometeoroids and orbital debris.

Propulsion

Mechanical components often need to be moved for deployment after launch or prior to landing. In addition to the use of motors, many one-time movements are controlled bypyrotechnic devices.

CONTROL

Robotic spacecraft use telemetry to radio back to Earth acquired data and vehicle status information. Although generally referred to as "remotely-controlled" or "telerobotic", the earliest orbital spacecraft – such as Sputnik 1 and Explorer 1 – did not receive control signals from Earth. Soon after these first spacecraft, command systems were developed to allow remote control from the ground. Increased autonomy is important for distant probes where the light travel time prevents rapid decision and control from Earth. Newer probes such as Cassini–Huygens and the Mars Exploration Rovers are highly autonomous and use on-board computers to operate independently for extended periods of time.

SPACE PROBES

A space probe is a scientific space exploration mission in which a spacecraft leaves Earth and explores space. It may approach the Moon, enter interplanetary, flyby or orbit other bodies, or approach interstellar space.

ROBOTIC SPACECRAFT SERVICE VEHICLES

- MDA Space Infrastructure Servicing vehicle — an in-space refueling depot and service spacecraft for communication satellites in geosynchronous orbit. Launch planned for 2015.
- Mission Extension Vehicle is an alternative approach that does not utilize in-space RCS fuel transfer. Rather, it would connect to the target satellite in the same way as MDA SIS, and then use "its own thrusters to supply attitude control for the target."

ROBOTIC TELESCOPE

Fig. "El Enano", a robotic telescope

A robotic telescope is an astronomical telescope and detector system that makes observations without the intervention of a human. In astronomical disciplines, a telescope qualifies as robotic if it makes those observations without being operated by a human, even if a human has to initiate the observations at the beginning of the night, or end them in the morning. A robotic telescope is distinct from a remote telescope, though an instrument can be both robotic and remote.

DESIGN

Robotic telescopes are complex systems that typically incorporate a number of subsystems. These subsystems include devices that provide telescope pointing capability, operation of the detector (typically a CCD camera), control of the dome or telescope enclosure, control over the telescope's focuser, detection of weather conditions, and other capabilities. Frequently these varying subsystems are presided over by a master control system, which is almost always a software component.

Robotic telescopes operate under closed loop or open loop principles. In an open loop system, a robotic telescope system points itself and collects its data without inspecting the results of its operations to ensure it is operating properly. An open loop telescope is sometimes said to be operating on faith, in that if something goes wrong, there is no way for the control system to detect it and compensate.

A closed loop system has the capability to evaluate its operations through redundant inputs to detect errors. A common such input would be position encoders on the telescope's axes of motion, or the capability of evaluating the system's images to ensure it was pointed at the correct field of view when they were exposed.

Most robotic telescopes are small telescopes. While large observatory instruments may be highly automated, few are operated without attendants.

HISTORY OF PROFESSIONAL ROBOTIC TELESCOPES

Robotic telescopes were first developed by astronomers after electromechanical interfaces to computers became common at observatories. Early examples were expensive, had limited capabilities, and included a large number of unique subsystems, both in hardware and software. This contributed to a lack of progress in the development of robotic telescopes early in their history.

By the early 1980s, with the availability of cheap computers, several viable robotic telescope projects were conceived, and a few were developed. The 1985 book,Microcomputer Control of Telescopes, by Mark Trueblood and Russell M. Genet, was a landmark engineering study in the field. One of this book's achievements was pointing out many reasons, some quite subtle, why telescopes could not be reliably pointed using only basic astronomical calculations. The concepts explored in this book share a common heritage with the telescope mount error modeling software called Tpoint, which emerged from the first generation of large automated telescopes in the 1970s, notably the 3.9m Anglo-Australian Telescope.

Since the late 1980s, the University of Iowa has been in the forefront of robotic telescope development on the professional side. The Automated Telescope Facility (ATF), developed in the early 1990s, was located on the roof of the physics building at the University of Iowa in Iowa City. They went on to complete the Iowa Robotic Observatory, a robotic and remote telescope at the private Winer Observatory in 1997. This system successfully observed variable stars and contributed observations to dozens of scientific papers. In May 2002, they completed the Rigel Telescope. The Rigel was a 0.37-meter (14.5-inch) F/14 built by Optical Mechanics, Inc. and controlled by the Talon program. Each of these was a progression toward a more automated and utilitarian observatory.

One of the largest current networks of robotic telescopes is RoboNet, operated by a consortium of UK universities. The Lincoln Near-Earth Asteroid Research (LINEAR) Project is another example of a professional robotic telescope. LINEAR's competitors, the Lowell Observatory Near-Earth-Object Search, Catalina Sky Survey, Spacewatch, and others, have also developed varying levels of automation.

In 2002, the RAPid Telescopes for Optical Response (RAPTOR) project pushed the envelope of automated robotic astronomy by becoming the first fully autonomous closed–loop robotic telescope. RAPTOR was designed in 2000 and began full deployment in 2002. Theproject was headed by Tom

Vestrand and his team: James Wren, Robert White, P. Wozniak, and Heath Davis. Its first light on one of the wide field instruments was in late 2001, with the second wide field system came online in late 2002. Closed loop operations began in 2003.

Originally the goal of RAPTOR was to develop a system of ground-based telescopes that would reliably respond to satellite triggers and more importantly, identify transients in real-time and generate alerts with source locations to enable follow-up observations with other, larger, telescopes. It has achieved both of these goals quite successfully. Now RAPTOR has been re-tuned to be the key hardware element of the Thinking Telescopes Technologies Project.

Its new mandate will be the monitoring of the night sky looking for interesting and anomalous behaviors in persistent sources using some of the most advanced robotic software ever deployed. The two wide field systems are a mosaic of CCD cameras. The mosaic covers and area of approximately 1500 square degrees to a depth of 12th magnitude. Centered in each wide field array is a single fovea system with a field of view of 4 degrees and depth of 16th magnitude.

The wide field systems are separated by a 38 km baseline. Supporting these wide field systems are two other operational telescopes. The first of these is a cataloging patrol instrument with a mosaic 16 square degree field of view down to 16 magnitude.

The other system is a.4m OTA with a yielding a depth of 19-20th magnitude and a coverage of.35 degrees. Three additional systems are currently undergoing development and testing and deployment will be staged over the next two years. All of the systems are mounted on custom manufactured, fast-slewing mounts capable of reaching any point in the sky in 3 seconds. The RAPTOR System is located on site at Los Alamos National Laboratory (USA) and has been supported through the Laboratory's Directed Research and Development funds.

In 2004, some professional robotic telescopes were characterized by a lack of design creativity and a reliance on closed source and proprietary software. The software is usually unique to the telescope it was designed for and cannot be used on any other system. Often, robotic telescope software developed at universities becomes impossible to maintain and ultimately obsolete because the graduate students who wrote it move on to new positions, and their institutions lose their knowledge. Large telescope consortia or government funded laboratories don't tend to have this same loss of developers as experienced by universities. Professional systems generally feature very high observing efficiency and reliability. There is also an increasing tendency to adopt ASCOM technology at a few professional facilities. The need for proprietary software is usually driven by the competition for research dollars between institutions.

HISTORY OF AMATEUR ROBOTIC TELESCOPES

In 2004, most robotic telescopes are in the hands of amateur astronomers. A prerequisite for the explosion of amateur robotic telescopes was the availability of relatively inexpensive CCD cameras, which appeared on the commercial market in the early 1990s. These cameras not only allowed amateur astronomers to make pleasing images of the night sky, but also encouraged more sophisticated amateurs to pursue research projects in cooperation with professional astronomers. The main motive behind the development of amateur robotic telescopes has been the tedium of making research-oriented astronomical observations, such as taking endlessly repetitive images of a variable star.

In 1998, Bob Denny conceived of a software interface standard for astronomical equipment, based on Microsoft's Component Object Model, which he called the Astronomy Common Object Model (ASCOM). He also wrote and published the first examples of this standard, in the form of commercial telescope control and image analysis programs, and several freeware components. He also convinced Doug George to incorporate ASCOM capability into a commercial camera control software program. Through this technology, a master control system that integrated these applications could easily be written in perl, VBScript, or JavaScript. A sample script of that nature was provided by Denny.

Following coverage of ASCOM in Sky & Telescope magazine several months later, ASCOM architects such as Bob Denny, Doug George, Tim Long, and others later influenced ASCOM into becoming a set of codified interface standards for freeware device drivers for telescopes, CCD cameras, telescope focusers, and astronomical observatory domes. As a result amateur robotic telescopes have become increasingly more sophisticated and reliable, while software costs have plunged. ASCOM has also been adopted for some professional robotic telescopes.

Meanwhile, ASCOM users designed ever more capable master control systems. Papers presented at the Minor Planet Amateur-Professional Workshops (MPAPW) in 1999, 2000, and 2001 and the International Amateur-Professional Photoelectric Photometry Conferences of 1998, 1999, 2000, 2001, 2002, and 2003 documented increasingly sophisticated master control systems. Some of the capabilities of these systems included automatic selection of observing targets, the ability to interrupt observing or rearrange observing schedules for targets of opportunity, automatic selection of guide stars, and sophisticated error detection and correction algorithms.

Remote telescope system development started in 1999, with first test runs on real telescope hardware in early 2000. RTS2 was primary intended for Gamma ray burst follow-up observations, so ability to interrupt observation was core part of its design. During development, it became an integrated observatory management suite. Other additions included use of the Postgresql database for storing targets and observation logs, ability to perform image processing including astrometry and performance of the real-time telescope corrections and a web-based user interface. RTS2 was from the beginning designed as a completely open source system, without any proprietary components. In order to support growing list of mounts, sensors, CCDs and roof systems, it uses own, text based communication protocol. The RTS2 system is described in papers appearing in 2004 and 2006.

The Instrument Neutral Distributed Interface (INDI) was started in 2003. In comparison to the Microsoft Windows centric ASCOM standard, INDI is a platform independent protocol developed by Elwood C. Downey of ClearSky Institute to support control, automation, data acquisition, and exchange among hardware devices and software frontends.

Significance

By 2004, robotic observations accounted for an overwhelming percentage of the published scientific information on asteroid orbits and discoveries, variable star studies,supernova light curves and discoveries, comet orbits and gravitational microlensing observations.

All early phase Gamma ray burst observations were carried by robotic telescopes.

SPACE PROBE

Fig. Pioneer H hangs in a museum

A space probe is a scientific space exploration mission in which a spacecraft leaves Earth and explores space. It may approach the Moon; enter interplanetary space; flyby, orbit or land on other planetary bodies; or approachinterstellar space. Space probes are a form of robotic spacecraft.

INTERPLANETARY TRAJECTORIES

Once a probe has left the vicinity of Earth, its trajectory will likely take it along an orbit around the Sun similar to the Earth's orbit. To reach another planet, the simplest practical method is a Hohmann transfer orbit. More complex techniques, such as gravitational slingshots, can be more fuel-efficient, though they may require the probe to spend more time in transit. Some high Delta-V missions (such as those with high inclination changes) can only be performed, within the limits of modern propulsion, using gravitational slingshots. A technique using very little propulsion, but requiring a considerable amount of time, is to follow a trajectory on the Interplanetary Transport Network.

SOME NOTABLE PROBES

Luna 16

First unmanned robotic sample return probe from the Moon.

Lunokhod 1

First rover on Moon. It landed on the Moon in 1970.

Mariner 10

First probe to Mercury.

Venera 3

Probe from the Soviet Union was the first man-made spacecraft to impact on another planet (Venus).

Venera 7

The Venera 7 probe was the first man-made spacecraft to successfully soft landing on another planet (Venus) and to transmit data from there back to Earth.

Mariner 9

Upon its arrival at Mars on November 13, 1971, Mariner 9 became the first space probe to maintain orbit around another planet.

Mars 3

First soft landing on Mars (between 1960 and 1973).

Sojourner

First successful rover on Mars.

Spirit and Opportunity

The Mars Exploration Rovers, Spirit and Opportunity surface and geology, and searched for clues to past water activity on Mars. They were each launched in 2003 and landed in 2004. Communication with Spirit stopped on sol 2210 (March 22, 2010). JPL continued to attempt to regain contact until May 24, 2011, when NASA announced that efforts to communicate with the unresponsive rover had ended. Opportunity arrived at Endeavour crater on 9 August 2011, at a landmark called Spirit Pointnamed after its rover twin, after traversing 13 miles from Victoria crater, over a three-year period. As of January 16, 2012,Opportunity has lasted for more than eight years on Mars — although the rovers were intended to last only three months.

Halley Armada

The first dedicated missions to a comet; in this case, to Halley's Comet during its 1985-86 journey through the inner solar system. It was also the first massive international coordination of space probes on an interplanetary mission, with probes specifically launched by the Soviet (now Russian) Space Agency, European Space Agency and Japan's ISAS (now integrated with NASDA to JAXA).

ICE

Original a solar observatory in the International Sun-Earth Explorer series, it was sent into solar orbit to make the first close observations of a comet, Comet Giacobini-Zinner, in 1985 as a prelude to studies of Halley's Comet.

Vega

Two Russian/French spacecraft. They dropped landers and balloons (first weather ballons deployed on another planet) at Venus before their rendezvous with Halley's Comet.

Sakigake

This Japanese probe was the first non-US, non-Soviet interplanetary probe.

Suisei

A second Japanese probe, it made ultraviolet wavelength observations of the comet.

Giotto

The first space probe to penetrate a comet's coma and take close-up images of its nucleus.

Genesis

First solar wind sample return probe from sun-earth L1.

Stardust

First sample return probe from comet tail.

NEAR Shoemaker

First probe to asteroid with landing.

Hayabusa

First sample return probe from asteroid.

Rosetta

The Rosetta space probe has flown by two asteroids and is aiming to rendezvous and explore comet 67P/Churyumov-Gerasimenko. It is scheduled to arrive at the comet in 2014.

Pioneer 10

First probe to Jupiter.

Pioneer 11

First probe to flyby two planets and first probe to Saturn.

Voyager 1

Voyager 1 is a 733-kilogram probe launched September 5, 1977. It is currently still operational, making it the longest-lasting mission of the U.S. National Aeronautics and Space Administration (NASA). It visited Jupiter and Saturn and was the first probe to provide detailed images of the moons of these planets. Voyager 1 is the farthest human-made object from Earth, traveling away from both the Earth and the Sun at a relatively faster speed than any other probe. As of September 12, 2013, Voyager 1 is about 12 billion miles from the Sun. On August 25, 2012, Voyager 1 became the first human made object to enter interstellar space. Voyager 1 has not had a functioning plasma sensor since 1980, but a solar flare in 2012 allowed scientists from NASA to measure vibrations of the plasma surrounding the craft. The vibrations allowed scientists to measure the plasma to be much denser than measurements taken in the far layers of our Heliosphere, thus concluding the craft had broken beyond the Heliopause.

Voyager 2

Voyager 2 was the first probe to complete the Planetary Grand Tour of the gas giants, and the first probe to visit Uranus and Neptune.Voyager 2 is the second-farthest human-made object from Earth, next to Voyager 1 at a distance of 101.2 AU as of July 11, 2013.

Huygens

First landing on Titan on January 14th 2005.

New Horizons

First probe to be launched to Pluto. Launched on 19 January 2006 with an estimated arrival date at the Pluto–Charon system of 14 July 2015.

Dawn

First spacecraft to visit and orbit a protoplanet (4 Vesta), entering orbit on July 16, 2011.

Juno

First probe to Jupiter without atomic battery, launched August 8, 2011.

Chang'e 2

First probe to orbit the Moon, visit Sun-Earth L2 Langrangian point and make a flyby of asteroid 4179 Toutatis.

Beyond the Solar System

Along with Pioneer 10, Pioneer 11, and its sister space probe Voyager 2, Voyager 1 is now an interstellar probe. Voyager 1 and Voyager 2 have both achieved solar escape velocity, meaning that their trajectories will not return them to the solar system.

UNMANNED RESUPPLY SPACECRAFT

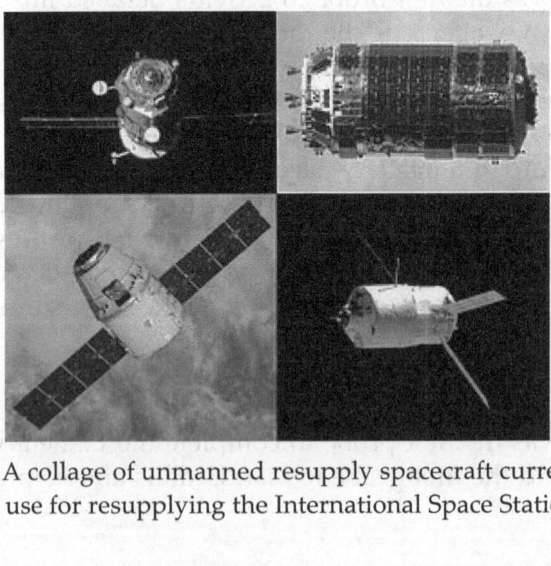

Fig. A collage of unmanned resupply spacecraft currently in use for resupplying the International Space Station

Unmanned resupply spacecraft are a special kind of robotic spacecraft that operate autonomously without a human crew, designed to support space station operation. This is different from space probes, whose missions are to conduct scientific investigations.

Resupply spacecraft have been used since 1978 and have serviced Salyut 6, Salyut 7, Mir and the International Space Station.

SPACECRAFT

Current spacecraft

- The Russian Progress spacecraft—developed by Russian Federal Space Agency
- The European Automated Transfer Vehicle—developed by the European Space Agency
- The Japanese H-II Transfer Vehicle—developed by the Japan Aerospace Exploration Agency
- The US-American Dragon spacecraft—developed by SpaceX, a private-sector company
- The US-American Cygnus spacecraft—developed by Orbital Sciences Corporation, a private-sector company

Spacecraft in development

- A Chinese cargo craft based on Tiangong-1 that will have a maximum diameter of 3.35 meters and a launch weight less than 13 tons.

CANCELED PROJECTS

- The American private-sector Kistler K-1 from Rocketplane Kistler, canceled October 2007 when failed to meet objectives.

SPACEX

Dragon

The American private space technology company that has made the most progress towards creating a spacecraft to deliver supplies to the International Space Station is also saying they will be the cheapest to operate. SpaceX will use their Falcon 9 rocket to launch the Dragon capsule into low Earth orbit. Using this combination SpaceX says they will be able to launch supplies or seven astronauts to the International Space Station for an average price of 57 million dollars per launch.

The United States government currently sends astronauts to the International Space Station using the Russian space agency at a cost of 63 million dollars a seat. The Falcon 9 rocket can also launch supplies into low

Earth orbit for approximately $2,500 per pound ($5,512/kg). SpaceX is also finishing the developmental stage on their Falcon Heavy rocket that is scheduled to make its first flight in 2014. The Falcon Heavy could launch supplies into low Earth orbit for a cost of $1,000 per pound ($2,205/kg). Those prices are less than half of the price that the United States Government spends now using the Russians and the Chinese. SpaceX believes they will able to save the United States government substantial costs using their rockets and capsules.

8

Spatial Language for Mobile Robots

INTRODUCTION

While all human languages can describe spatial representations, people speaking different languages will use different frames of reference: intrinsic (from the point of view of the object), relative (from the point of view of the speaker or some other viewer) or absolute (e.g. North, South, East and West) (Levinson, 1996). These frames of reference can be used to construct or describe spatial relationships in the world. The use of different frames of reference in different languages indicates that language may restructure the spatial representations of the language speaker, rather than the existence of innate and universal spatial concepts (Majid, Bowerman, Kita, Haun, & Levinson, 2004).

Computational modeling of language evolution provides a means of investigating ontology, grounding, learnability, and generalization in languages that evolve in populations of agents. The use of simulation techniques can add to the debate on the origins and evolution of language by determining factors that are important for evolving communication systems. Language games are a possible framework for language models in which agents engage in tasks requiring communication.

These games have been used to evolve lexicons (Hutchins & Hazlehurst, 1995), categories (Cangelosi & Harnad, 2001), and grammars (Batali, 2002) in populations of agents.

The symbol grounding problem (Harnad, 1990) is a major issue for computational models of language. Without the grounding of meanings in the world, symbols refer only to other symbols with no association between the symbols and the world. One way to address the symbol grounding problem in computational models of language is to conduct language research with real or simulated robots (Marocco, Cangelosi, & Nolfi, 2003; Roy, 2001; Steels, 1999; Vogt, 2000).

In robot language research, the environments are often simplified and idealized compared to the real world. In the Talking Heads Experiment (Steels,

1999) geometric shapes were used rather than 'real world' objects such as tables and chairs. The languages evolved in the Talking Heads Experiment used a relative frame of reference to talk about the different shapes in the scene using meanings such as 'left' and 'right'.

One way to extend robot language research is to use mobile robots that interact with a real world environment, using navigation systems to build up internal maps of the world. The use of mobile autonomous agents that move in a real environment enables the evolution of spatial languages using both relative and absolute frames of reference. The visual input of the robot would be used in a relative frame of reference, where the scenes can be categorized with respect to what the world looks like from the perspective of the robot. The internal maps would be used in an absolute frame of reference. The languages evolved could provide a methodology to investigate the structure of languages that describe space.

This paper introduces RatChat, a project that uses RatSLAM, an established mobile robot platform, to develop a framework for the robots to evolve a language describing their environment. The RatChat and RatSLAM projects are described in Section 2. A study using this platform to evolve spatial languages is presented in Section 3, followed by a general discussion and conclusion.

RATCHAT

Simultaneous Localisation and Mapping (SLAM) is a methodology for robot map building and navigation. RatSLAM is a model of SLAM, based on the hippocampal complex in rodents, that uses a combination of the properties of grid based, topological, and landmark representations to keep a sense of space while adding robustness and adaptability (Milford, Wyeth, & Prasser, 2004).

The inputs to the RatSLAM system include odometry and vision with the resulting map represented by pose cells. Active pose cells represent the current location and orientation of the robot, and are arranged in (x,y,?) for ease of visualization. With RatSLAM, robots use the appearance of an image to aid localization by learning to associate the appearance of a scene and its position estimate (Prasser, Wyeth, & Milford, 2004).

RatChat aims to evolve a shared lexicon between robots grounded in perceptions, local views, and behaviors using a language game framework. The evolution of languages for locations will be explored, later extending the vocabulary of the robots to include objects. The challenge is for the robots to categorize their internal representations and label these with appropriate generalization and variability. The shared lexicon should allow the robots to agree on words for categories while including sufficient diversity for different categories to have different labels. As the language is expanded to include objects, more emphasis will be on the visual inputs of the robots.

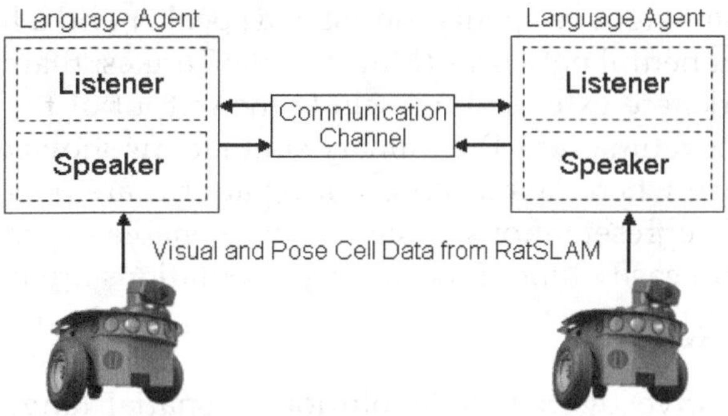

Fig. The framework for a language game. Each language agent obtains visual and pose cell data from the RatSLAM system. A communication channel is set up between the agents, allowing the speaker for each agent to produce utterances, and the listener for each agent to receive utterances for comprehension.

Fig. The robot's world comprises halls and open plan offices. A simulated world has been built to mirror the real world. The features of the environment shown in the visual images seen by the robot include the floor, walls, desks, chairs, and filing cabinets. The left image is from the robot's camera and the right image is the same location in the simulated world.

The RatChat language agents consist of a speaker and a listener based on simple recurrent neural networks (Elman, 1990; Tonkes, Blair, & Wiles, 2000). Speaker networks are extended to include the output of the network in the context for the next time step. Preliminary simulations showed that languages are easier to learn when the meaning space patterns are non-orthogonal and that distributed representations in signal space enable expressive languages to be found more easily than if localist representations are used.

A SPATIAL LANGUAGE

This study investigated the evolution of spatial languages using the visual and pose cell representations of the robot, looking at the expressivity of the languages evolved, and how the languages categorized the world of the robot.

Methods: The visual input for this study was every 100th scene in a series of 10000 visual scenes of 12x8 gray scale arrays obtained from a run of the robot in the simulated world. The pose cell input for the study was every 100th pattern in a series of 10000 pose cell patterns from the same run. The number of cells was reduced from 440640 to 610 by reducing the resolution of the pose cells (4x4x4 pose cells to 1 pose cell), and by discarding cells that are inactive in every pattern. For a third representation, the pose cells were processed using a hybrid system based on Self Organizing Maps (SOMs) (Kohonen, 1995). In the processing system, a SOM was trained on the input series for 1000 epochs. The output of the SOM was a 12x8 set of competitive units organized in a hexagonal pattern. To construct a distributed activation the actual output values of the units were converted to values between 0 and 1.

For the signal representation, utterances consisted of a sequence of three syllables. Each syllable was represented by a ten unit binary vector in which the two most active units were set to one, with all other units set to zero.

One way to measure understanding is to test how well an agent has categorized the world. The representations of the world are presented to the speaker, resulting in words associated with each pattern. Listeners produce a prototype for each unique utterance. If the original input pattern presented to the speaker is closest to the prototype for the utterance used by the speaker, this pattern has been correctly categorized. When many of the patterns are associated with one word, the agents will categorize more patterns correctly, but the language does not divide the meaning space effectively. A more appropriate measure of understanding is the number of patterns correctly categorized divided by the largest category size, indicating how well the language divides up the meaning space, and how well the agent understands the language.

In this study, ten agents were evolved individually for 100 generations to produce languages based on each set of inputs (vision, pose cells and processed pose cells). A simple (1+1)-evolutionary strategy (Beyer & Schwefel, 2002) was used to evolve the agent's speaker, introducing variability in the language. At each step, the agent's speaker was evolved and the agent's listener was trained on the language from the speaker for 500 epochs using the Back Propagation Through Time algorithm (Rumelhart, Widrow, & Lehr, 1994). The agents were evaluated with a fitness function based on the measure of understanding described above. If the listener trained on the mutant languages were better at categorizing the input patterns than the listener trained on the current champion language, then the mutant became the champion. The languages produced by the agents for each set of inputs were compared for expressiveness, categorization and how the meaning space was divided.

Results: The agents evolved with visual scenes as inputs produced languages with an average of 24.2 words. The average number of scenes correctly categorized by the agents was 53.4 out of 100. One highly expressive language had 67 unique words of which 47 were associated with single scenes. Words often appeared to group several different types of images together, with the resulting prototype visual scene for the word a combination of these scenes. One set of similar scenes were those in which the robot faced a white wall with a strip of black next to the floor. All of the languages other than the most expressive language grouped together some of these scenes.

Table. Properties of the languages evolved with different sets of input

	Number of Unique Words (avg (std))	**Number of Patterns Correctly Categorized (avg (std))**
Vision	24.2 (17.3)	53.4 (13.5)
Pose Cells	23.2 (12.4)	22.6 (10.4)
Processed Pose Cells	10.9 (6.4)	58.7 (10.4)

The agents evolved with pose cells as inputs produced languages with an average of 23.2 words. The average number of scenes correctly categorized by the agents was 22.6 out of 100. The majority of the words were associated with single input patterns or a small number of input patterns, scattered across the space. Some words group together input patterns that are close together in space, but these words are also generally associated with a small number of input patterns from other areas.

The agents evolved with processed pose cells as inputs produced languages with an average of 10.9 words. The average number of scenes correctly categorized by the agents was 58.7 out of 100. These languages had less words associated with single input patterns and more words associated with many input patterns spread across the entire space. The larger languages had more words associated with groups of input patterns that were close together in space.

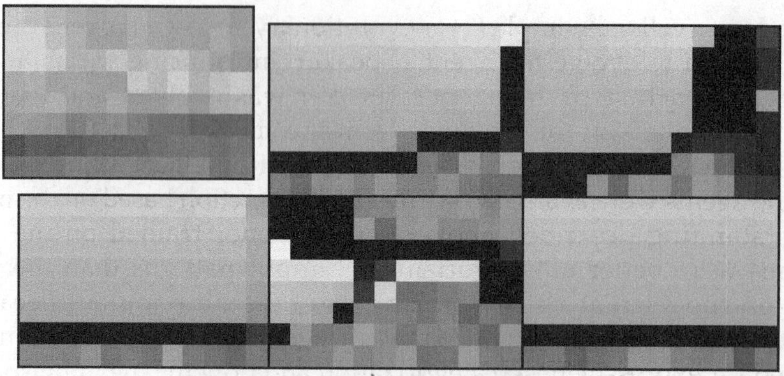

Fig. The prototype for the word 'kufufu' (top left) and the five scenes that are associated with this word in a language with 27 unique words. Most of the scenes associated with 'kufufu' show a white wall with a black strip, although the bottom middle scene has different features.

Discussion: Expressivity is an important feature of language, where unique words are used for unique meanings. In this simulation, expressivity is indicated by the number of unique words. The vision and pose cell representations resulted in languages with an average of over 20 unique words for the 100 input patterns, while the processed pose cell representation resulted in languages with an average of 10.9 unique words. This reduction in expressivity for the processed pose cell representation indicates that the unique information in the input patterns may be lost when the pose cell representation is processed.

The number of categories correct indicates how well the language categorizes the world. The processed pose cell languages were most successful at clustering input patterns that were close together in space, with distinct clusters associated with single words. The unprocessed pose cell languages were not as successful at categorizing the patterns, which may be due to the size and sparseness of the pose cell representation, and can be addressed by processing the pose cell representation.

Some of the agents using languages evolved with vision were successful in grouping together similar scenes, however many of the words in the vision languages grouped together images that were dissimilar as well as similar, or were associated with single images. In this study, raw vision as an input provided a structure that allowed some languages to evolve to successfully categorize the world. Processing the scenes prior to the language agent may extract the important information from each scene that is necessary for languages to consistently evolve with expressivity and categorization.

General discussion and conclusion

The RatChat project aims to explore the structure of languages that describe space using mobile robots. The simulations presented in this paper represent agents developing their internal representations of the world prior to playing naming games in populations of agents, and have provided insight into the expressivity, categorization, and structure of languages that can evolve from visual and pose cell representations.

There is a tradeoff between expressivity, with unique words for unique meanings, and categorization, with the use of one word for a group of similar meanings. The degree of expressivity and categorization can be altered by processing the inputs, as can be seen with the pose cell representation: the unprocessed languages are more expressive, while the processed languages are better at categorizing the world.

We are currently running simulations to scale up these results with further studies into processing the robot representations prior to the language networks and evolving languages in populations of agents.

CHALLENGES IN BUILDING ROBOTS THAT IMITATE PEOPLE

Humans (and some other animals) acquire new skills socially through direct tutelage, observational conditioning, goal emulation, imitation, and other methods (Galef, 1988; Hauser, 1996). These social learning skills provide a powerful mechanism for an observer to acquire behaviors and knowledge from a skilled individual (the model). In particular, imitation is an extremely powerful mechanism for social learning which has received a great deal of interest from researchers in the fields of animal behavior and child development.

Similarly, social interaction can be a powerful way for transferring important skills, tasks, and information to a robot. A socially competent robot could take advantage of the same sorts of social learning and teaching scenarios that humans readily use. From an engineering perspective, a robot that could imitate the actions of a human would provide a simple and effective means for the human to specify a task to the robot and for the robot to acquire new skills without any additional programming.

From a computer science perspective, imitation provides a means for biasing interaction and constraining the search space for learning. From a developmental psychology perspective, building systems that learn through imitation allows us to investigate a minimal set of competencies necessary for social learning. We can further speculate that constructing an artificial system may provide useful information about the nature of imitative skills in humans (or other animals).

Initial studies of social learning in robotics focused on allowing one robot to follow a second robot using simple perception (proximity and infrared sensors) through mazes (Hayes & Demiris, 1994) or an unknown landscape (Dautenhahn, 1995). Other work in social learning for autonomous robots addressed learning inter-personal communication protocols between similar robots (Steels, 1996), and between robots with similar morphology but which differ in scale (Billard & Dautenhahn, 1998). Robotics research has also focused on how sequences of known behaviors can be chained together based on input from a model.

Mataric, Williamson, Demiris & Mohan (1998) used a simulated humanoid to learn a sequence of gestures from a set of joint angles recorded from a human performing those same gestures, and Gaussier, Moga, Banquet, and Quoy (1998) used a neural network architecture to allow a robot to sequence motor primitives in order to follow the trajectory of a teacher robot. One research program has addressed how perceptual states can be categorized by matching against models of known behaviors; Demiris and Hayes (1999) implemented an architecture for the imitation of movement on a simulated humanoid by predictively matching observed sequences to known behaviors. Finally, a variety of research programs have aimed at training robots to perform single tasks by observing a human demonstrator.

Schaal (1997) used a robot arm to learn a pendulum balancing task from constrained visual feedback, and Kuniyoshi, Inaba, and Inoue (1994) discussed a methodology for allowing a robot in a highly constrained environment to replicate a block stacking task performed by a human but in a different part of the workspace.

Traditionally in robot social learning, the model is indifferent to the attempts of the observer to imitate it. In general, learning in adversarial or indifferent conditions is a very difficult problem that requires the observer to decide who to imitate, what to imitate, how to imitate, and when imitation is successful.

To make the problem tractable in an indifferent environment, researchers have vastly simplified one or more aspects of the environment and the behaviors of the observer and the model. Many have simplified the problem by using only simple perceptions which are matched to relevant aspects of the task, such as Kuniyoshi, Inaba, and Inoue's (1994) use of white objects on a black background without any distractors or Mataric, Williamson, Demiris, and Mohan's (1998) placement of reflective markers on the human's joints and use of multiple calibrated infrared cameras.

Others have assumed the presence of a single model which is always detectable in the scene and which is always performing the task that the observer is programmed to learn, such as Gaussier, Moga, Banquet, and Quoy (1998), and Schaal (1997). Many have simplified the problem of action selection by having limited observable behaviors and limited responses (such as Steels (1996) and Demiris and Hayes (1999)), by assuming that it is always an appropriate time and place to imitate (such as Dautenhahn (1995)), and by fixing the mapping between observed behaviors and response actions (such as Billard & Dautenhahn (1998)).

Few have addressed the issue of evaluating the success of an imitative response; most systems use a single, fixed success criteria which can only be used to learn a strictly specified task with no hope for error recovery (although see Nehaniv and Dautenhahn (1998) for one treatment of evaluation and body mapping).

Our approach is to constrain the learning scenario in a different manner – we assume that the model is motivated to help the observer learn the task. A good teacher is very perceptive to the limitations of the learner and sets the complexity of the instruction and task accordingly. As the learner's performance improves, the instructor incrementally increases the complexity of the task. In this way, the learner is always competent but slightly challenged – a condition amenable for successful learning. This assumption allows us to build useful implementations on our robots, but limits the applicability of these results to less constrained learning environments (such as having an indifferent model). However, we believe that the problems that must be addressed in building systems with the assumption of an active instructor are also applicable to robotics programs that use other assumptions and to investigations of social learning in natural systems.

We will use the word imitate to imply that the observer is not merely replicating the actions of the model but rather is attempting to achieve the goal of the model's action by performing a novel action similar to that observed in the model. Although we focus on this relatively strong definition, more basic forms of social learning share many of the same challenges. Simpler mechanisms such as stimulus enhancement, emulation, and mimicry must also address challenges such as determining what actions are relevant in the scene and finding conspecifics, while other challenges (such as determining the goal behind an action) are specific to this definition of imitation. It is an open question as to whether or not inferring intent is necessary to explain particular behaviors (Byrne, 1999). However, for a robot to fulfill the expectations of a human instructor, the robot must have a deeper understanding of the goal and intent of the task it is learning to perform.

In this chapter, we outline four hard problems in building robots that imitate people and discuss how the social cues that humans naturally and intuitively provide could be used by a robot to solve these difficult problems. By attempting to build systems that imitate, we are forced to address issues which are not currently discussed in developmental psychology, animal behavior, or other research domains. However, we believe that these issues must be addressed by any creature or artifact that learns through imitation, and the study of these issues will yield greater insight into natural systems. We will present our progress towards implementing a set of critical social skills on two anthropomorphic robots, and discuss initial experiments which use these skills to benefit the imitative learning process.

HARD PROBLEMS IN ROBOT IMITATION

The ability to imitate relies upon many perceptual, cognitive, and motor capabilities. Many of these requirements are precursor skills which are necessary before attempting any task of this complexity, but which are not directly related to the act of imitation. For example, the robot will require systems for basic visual-motor behaviors (such as smooth pursuit tracking and vergence), perceptual abilities for detecting motion, color, and scene segmentation, postural control, manipulative abilities such as reaching for a visual target or controlled-force grasping, social skills such as turn taking and recognition of emotional states, as well as an intuitive physics (including object permanence, support relations, and the ability to predict outcomes before attempting an action).

Even if we were to construct a system which had all of the requisite precursor skills, the act of imitation also presents its own unique set of research questions. Each of these questions is a complex research problem which the robotics community has only begun to address. In this chapter, we focus on four of these questions:

- How does the robot know when to imitate?
- How does the robot know what to imitate?
- How does the robot map observed actions into behavioral responses?
- How does the robot evaluate its actions, correct errors, and recognize when it has achieved its goal?

To investigate these questions, consider the following example:

- The robot is observing a model opening a glass jar. The model approaches the robot and places the jar on a table near the robot. The model rubs his hands together and then sets himself to removing the lid from the jar. He grasps the glass jar in one hand and the lid in the other and begins to unscrew the lid. While he is opening the jar, he pauses to wipe his brow, and glances at the robot to see what it is doing. He then resumes opening the jar. The robot then attempts to imitate the action.

How does the robot know when to imitate?

A socially intelligent robot should be able to use imitation for the variety of purposes that humans do. Human children use imitation not only to acquire new skills, but also to acquire new goals from their parents. By inferring the intention behind the observed actions, children can gain an understanding of the goals of an individual.

Children also use imitation to acquire knowledge about socializing, including the social conventions of their culture and the acceptable dynamics necessary for social communication. Imitation can be a mechanism for developing social attachments through imitative play and for gaining an understanding of people. Just as infants learn about physical objects by acting on them, infants learn about people by interacting with them. As Meltzoff and Moore (1994) wrote, "Imitation is to understanding people as physical manipulation is to understanding things." Imitation can also be used to explore and expand the range of possible actions in the child's repertoire, learning new ways of manipulating objects or new motor patterns that the child might not otherwise discover. Finally, imitation can be a mechanism for establishing personal identity and discovering distinctions between self and other. Meltzoff and Moore (1994) have proposed that deferred imitation may serve to establish the identity of a previously encountered individual.

A social robot should selectively use imitation to achieve many of these goals. However, the robot must not merely be a "puppet on a string." The robot must decide whether or not it is appropriate to engage in imitative behavior based on the current social context, the availability of a good model, and the robot's internal goals and motivations. For example, the robot may need to choose between attending to a learning opportunity or fulfilling another goal, such as recharging its batteries.

This decision will be based upon the social environment, how likely the robot is to have another opportunity to engage in that particular learning opportunity, the current level of necessity for charging the batteries, the quality of the instruction, and other competing motivations and goals. Furthermore, the robot should also recognize when imitation is a viable solution and act to bring about the social context in which it can learn by observation, perhaps by seeking out an instructor or motivating the instructor to perform a certain task.

How does the robot know what to imitate?

Faced with an incoming stream of sensory data, the robot must make a number of decisions to determine what actions in the world are appropriate to imitate. The robot must first determine which agents in the scene are good models (and be able to avoid bad models). The robot must not only be able to distinguish the class of stimuli (including humans and perhaps other robots) which might be a good model but also determine if the current actions of that agent are worthy of imitation. Not all humans at all times will be good models, and imitation may only be appropriate under certain circumstances.

Once a model has been selected, how does the robot determine which of the model's actions are relevant to the task, which may be part of the social/ instructional process, and which are circumstantial? In the example above, the robot must segment the scene into salient objects (such as the instructor's hand, the lid, and the jar) and actions (the instructor's moving hand twisting the cap and the instructor's head turning toward the robot).

The robot must determine which of these objects and events are necessary to the task at hand (such as the jar and the movement of the instructor's elbow), which events and actions are important to the instructional process but not to the task itself (such as the movement of the instructor's head), and which are inconsequential (such as the instructor wiping his brow). The robot must also determine to what extent each action must be imitated. For example, in removing the lid from a jar, the movement of the instructor's hand is a critical part of the task while the instructor's posture is not The robot must also recognize the important aspects of the objects being manipulated so that the learned action will be applied to only appropriate objects of the same class (Scassellati, 1999B).

How does the robot map observed actions into behavioral responses?

Once the robot has identified salient aspects of the scene, how does it determine what actions it should take? When the robot observes a model opening a jar, how does the robot convert that perception into a sequence of motor actions that will bring its arm to achieve the same result? Mapping from one body to another involves not only determining which body parts have similar structure but also transforming the observed movements into motions that the robot is capable of performing. For example, if the instructor is unscrewing the lid of the jar, the robot must first identify that the motion of the arm and hand are relevant to the task and determine that its own hand and arm are capable of performing this action. The robot must then observe the movements of the instructor's hand and arm and map those movements into the motor coordinates of its own body.

How does the robot evaluate its actions, correct errors, and recognize success?

Once a robot can observe an action and attempt to imitate it, how can the robot determine whether or not it has been successful? In order to compare its actions with respect to those of the model, the robot must be able to identify the desired outcome and to judge how similar its own actions were to that outcome. If the robot is attempting to unscrew the lid of a jar, has the robot been successful if it merely mimics the model and rotates the lid but leaves the lid on the jar? Is the robot successful if it removes the lid by pulling instead of twisting? Is the robot successful if it smashes the jar in order to open it? In the absence of internal motivations that provide feedback on the success of the action, the evaluation will depend on an understanding of the goals and intentions of the model. Further, if the robot has been unsuccessful, how does it determine which parts of its performance were inadequate? The robot must be able to diagnose its own errors in order to incrementally improve performance.

APPROACH

Our approach to building systems that address the problems of determining saliency and relevance, mapping observed actions into behavioral responses, and implementing incremental refinement focuses on three keystones. First, saliency results from a combination of inherent object qualities, contextual influences, and the model's attention. This provides the basis for building perceptual systems that can respond to complex social situations. Second, our robots utilize similar physical morphologies to simplify the task of body mapping and recognizing success. By building human-like robots we can vastly simplify the problems of mapping perceived actions to behavioral responses while providing an interface that is intuitive and easy to correct.

Third, our systems exploit the structure of social interactions. By recognizing the social context and the stereotypical social actions made by the model, our robots can recognize saliency. By engaging in those same types of stereotypical social actions, the dynamics between the robot and the model provide a simplified means for recognizing success and diagnosing failures.

Saliency results from a combination of inherent object qualities, contextual influences, and the model's attention

Knowing what to imitate is fundamentally a problem of determining saliency. Objects can gain saliency (that is, they become the target of attention) through a variety of means. At times, objects are salient because of their inherent properties; objects that move quickly, objects that have bright colors, and objects that are shaped like faces are all likely to attract attention. (We call these properties inherent rather than intrinsic because they are perceptual properties, and thus are observer-dependant and not strictly a quality of an external object.) Objects can also become salient through contextual effects. The current motivational state, emotional state, and knowledge of the observer can impact saliency.

For example, when the observer is hungry, images of food will have higher saliency than they otherwise would. Objects can also become salient if they are the focus of the model's attention. For example, if the model is staring intently at a glass jar, the jar may become a salient part of the scene even if it is otherwise uninteresting. Fundamental social cues (such as gaze direction) can also be used by the observer to determine the important features of a task. People naturally attend to the key aspects of a task while performing that task. For example, when opening the jar, the model will naturally look at the lid as he grasps it and at his own hand while twisting off the lid. By directing its own attention to the object of the model's attention, the observer will automatically attend to the critical aspects of the task. In the case of social instruction, the observer's gaze direction can also serve as an important feedback signal for the instructor. For example, if the observer is not attending to the jar, then the instructor can actively direct the observer's attention by increasing the jar's saliency, perhaps by pointing to it or tapping on it.

Utilize similar physical morphologies

Three of the problems outlined above can be simplified by assuming a similar physical morphology between the model and the observer. If the observer and model have a similar shape, the perceptual task of determining saliency can be constrained by the possible actions of the observer. If the observer witnesses an ambiguous motion of the model's arm, the observer can postulate that the perception must have been one of the actions which it could possibly perform in that situation and eliminate any other possible perceptual interpretations.

The mapping problem can also be simplified by having similar physical morphologies. If the observer can identify that it is the model's arm that is moving, it need not initially try to match that motion with an action that it is capable of performing only with its mouth or legs. Additionally, the position of the model's arm serves as a guideline for an initial configuration for the observer's arm.

A different morphology would imply the need to solve an inverse kinematics problem in order to arrive at a starting position or the more complicated problem of mapping unlike body parts between model and observer (for example, see the chapter by Hermann for imitation between dolphins and humans). In general this transformation has many solutions, and it is difficult to add other constraints which may be important (e.g., reducing loading or avoiding obstacles). By constraining the space of possible mappings, the computational complexity of the task is reduced.

Similar physical morphology also allows for a more accurate evaluation. If the observer's morphology is similar to the model's, then the observer is likely to have similar failure modes. This potentially allows the observer to characterize its own failures by observing the failures of the model. If the observer watches the model having difficulty opening the jar when his elbows are close together, the observer may be able to extrapolate that it too will fail without sufficient leverage.

In situations where the model is taking an active role in instructing the observer, a similar morphology also allows the model to more easily identify and correct errors from the observer. If the observer's arms are too close together when attempting to open the jar, the model's knowledge about his own body will assist him in evaluating the failure mode and in providing an appropriate solution.

Exploit the structure of social interactions

Social interactions have structure that can be exploited to simplify the problems of imitation. By recognizing the appropriate social context, the observer can limit the number of possible perceptual states and determine whether the attention state of the model is an appropriate saliency signal. When the model is performing a manipulative task, the focus of attention is often very relevant. However, when engaged in some social contexts, the focus of attention is not necessarily important. For example, it is customary in many cultures to avert eye contact while taking one's turn in a conversation and to establish eye contact when ending a turn. Exploiting these rules of social conduct can help the observer to recognize the possible value of the attention state of the model (thus simplifying the saliency problem).

The structure of social interactions can also be used to provide feedback in order to recognize success and correct failures. In the case of social instruction, the difficulty of obtaining success criteria can be simplified by exploiting the natural structure of social interactions. As the observer acts, the facial expressions (smiles or frowns), vocalizations, gestures (nodding or shaking of the head), and other actions of the model all provide feedback that will allow the observer to determine whether or not it has achieved the desired goal.

The structure of instructional situations is iterative; the instructor demonstrates, the student performs, and then the instructor demonstrates again, often exaggerating or focusing on aspects of the task that were not performed successfully. The instructor continually modifies the way he performs the task, perhaps exaggerating those aspects that the student performed inadequately, in an effort to refine the student's subsequent performance. By repeatedly responding to the same social cues that initially allowed the observer to understand and identify which salient aspects of the scene to imitate, the observer can incrementally refine its approximation of the actions of the instructor.

Monitoring the structure of the social interaction can assist the instructor in maintaining an appropriate environment for learning. Expressive cues such as facial expressions or vocalizations can regulate the rate and quality of instruction. The instructor modifies both the speed and the content of the demonstration based on feedback from the student. By appearing confused, the student causes the instructor to slow down and simplify the demonstration.

Recognizing the appropriate social context can be an important cue in knowing when imitation is an appropriate solution to a problem. Internal motivations will serve as a primary mechanism for determining when to search for an appropriate model and when an attempt to perform an imitative act is appropriate. However, opportunistic use of good models in the environment can also be important in learning new skills. By recognizing which social contexts are likely to produce a good model behavior, the robot can exploit learning opportunities when they arise.

ROBOTIC IMPLEMENTATIONS

For the past four years, our group at the MIT Artificial Intelligence Laboratory has been attempting to build anthropomorphic robots that respond socially (Brooks et al., 1998). Building a system that can imitate requires the integration of many different social, perceptual, cognitive, and motor skills. To date, we have constructed some modules which will be useful components of a social learning mechanism. We still require many additional components, and we have yet to meet the challenge of integrating all of these components into a system that can learn from a human instructor.

In this section, we will describe some of the components which have already been implemented to address a few of the problems of social interaction, including a perceptual system for finding the model using face detection and skin color detection, a context-sensitive attention system, a system for producing expressive displays through facial expressions and body posture, and a system for regulating social exchanges to optimize the learning environment. Each of these components has been evaluated individually using traditional engineering techniques. In some cases, it is appropriate to compare the performance of a module with humans or animal data. Once all of the necessary components are integrated, we can ultimately evaluate the complete system using the same techniques that are used to characterize human behavior. Because the robot is embodied in the world, it can be evaluated side-by-side against a human in the same physical environment and in the same social context (using the same instructor and the same task). We begin with a description of the two robot platforms.

Robot Platforms

Our work with imitation has focused on two robot platforms: an upper-torso humanoid robot called Cog and an active vision system enhanced with facial features called Kismet. A basic repertoire of perceptual capabilities and sensory-motor skills have been implemented on these robots.

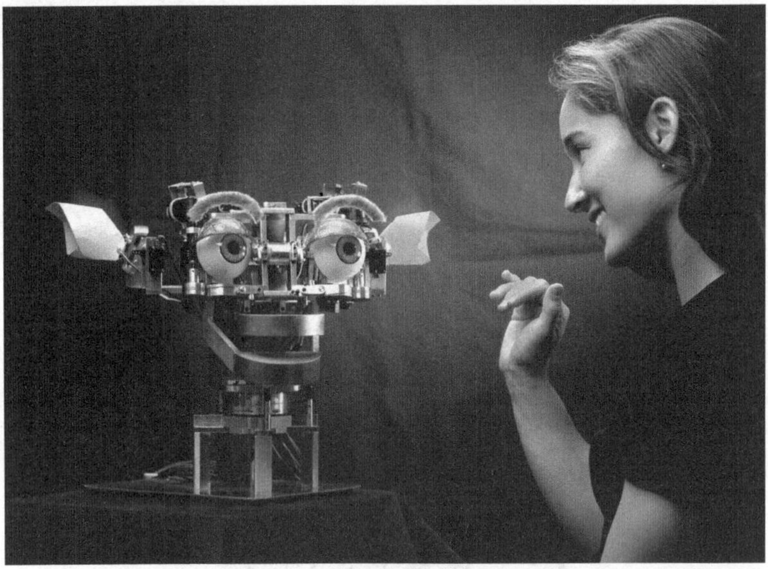

Fig. Cog (left) and Kismet (right), our two anthropomorphic robot platforms.

Cog approximates a human being from the waist up with twenty-two degrees-of-freedom (DOF) and a variety of sensory systems. The physical structure of the robot, with movable torso, arms, neck and eyes gives it human-like motion, while the sensory systems (visual, auditory, vestibular, and proprioceptive) provide rich information about the robot and its immediate environment. The robot Kismet is based on the same active vision system used on Cog. Kismet has an additional fifteen degrees-of-freedom in facial expressions, including eyebrows that lift and arch, ears that lift and rotate, eyelids, lips, and a mouth. The robot is able to show a wide variety of facial expressions and displays which it uses to engage a human in face-to-face exchanges (Breazeal & Scassellati, 1999a).

By focusing on robotic platforms that are anthropomorphic, we simplify the problems of social interaction in three ways. First, it allows for a simple and natural means of interaction. People already know how to provide the robot with appropriate feedback, how to attract its attention, and can guess what capabilities it might possess. Second, the responses of the robot can be easily identified and interpreted by a naive observer. Third, by having a similar body structure, the problem of mapping observed actions onto the robot's own body is simplified.

Finding a good model using face detection and skin color detection

For our robots, one of the first tasks that must be performed is locating an appropriate model. Because we assume that a good model will attempt to assist the robot and because human instructors attend to their students throughout the instructional process, the robot should be most interested in human faces which are oriented toward it. Difficulties with occlusion and multiple-viewpoint recognition can be avoided because a helpful instructor will position himself in face-to-face contact with the robot.

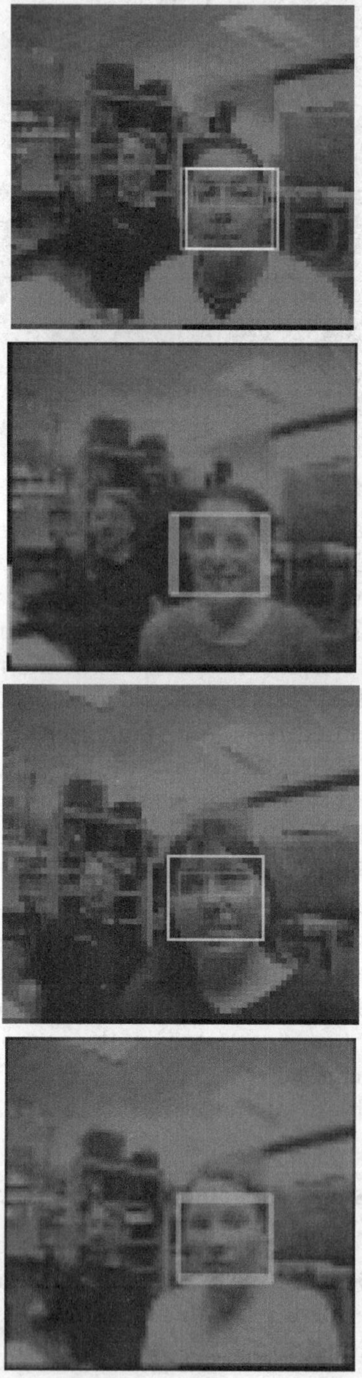

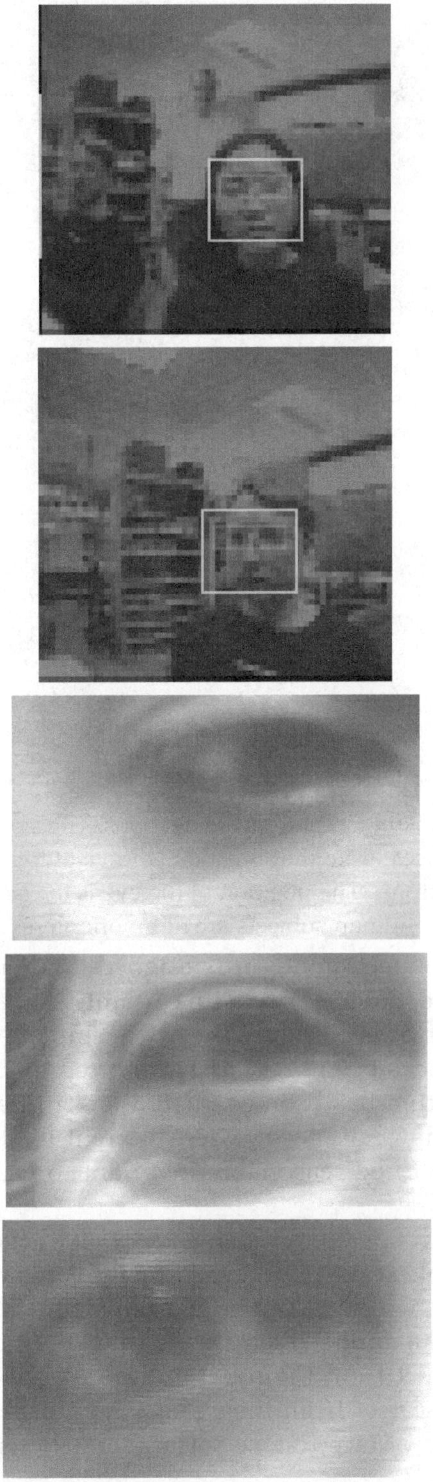

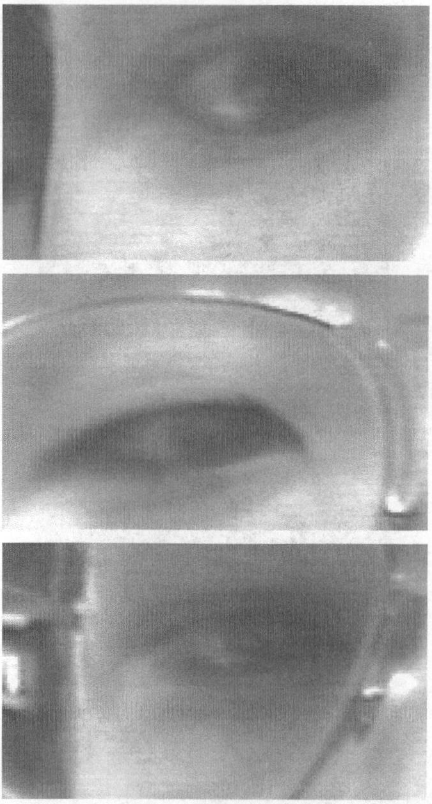

Fig. Examples of successful face and eye detections. The system locates faces in the peripheral camera, saccades to that position (shown at top), and then extracts an image of the eye (bottom). The position of the eye is inexact, in part because the human subjects are not motionless.

Our face detection techniques are designed to identify locations that are likely to contain a face, not to verify with certainty that a face is present in the image. The face detector is based on the ratio-template technique developed by Sinha (1996), and has been previously reported by Scassellati (1998). The ratio template algorithm has been evaluated on Turk and Pentland's (1991) database of frontal views of faces under different lighting and orientations, and has been shown to be reasonably invariant to changes in illumination and rotation.

The algorithm can operate on each level of an image pyramid in order to detect faces at multiple scales. In the current implementation, due to limited processing capability, we elected to process only a few image scales for faces. A 14x16 ratio template applied to a 128x128 image finds faces in a range of approximately 6-15 feet from the robot and applied to a 64x64 image finds faces in a range of 3-6 feet from the robot. This range was suitable for our current investigations of face-to-face social interactions, and could easily be expanded with additional processors. The implemented face detector operates at approximately 20 Hz. In combination with this template-based method, we also use a filter that selects skin-color regions from the image by selecting pixel locations that fall within a pre-specified range in the color space. These two techniques allow us to recognize the location of potential models.

A context-dependant attention system for determining saliency

To recognize salient objects, we have been constructing attention and perception systems that combine information on visual motion, innate perceptual classifiers such as face detectors, color saliency, depth segmentation, and auditory information with a habituation mechanism and a motivational and behavioral model. This attention system allows the robot to selectively direct computational resources and exploratory behaviors toward objects in the environment that have inherent or contextual saliency.

Title:
Microsoft PowerPoint - attn.ppt
Creator:
PSCRIPT.DRV Version 4.0
Preview:
This EPS picture was not saved
with a preview included in it.
Comment:
This EPS picture will print to a
PostScript printer, but not to
other types of printers.

Fig. Overview of the attention system. A variety of visual feature detectors (color, motion, and face detectors) combine with a habituation function to produce an attention activation map. The attention process influences eye control and the robot's internal motivational and behavioral state, which in turn influence the weighted combination of the feature maps. Displayed images were captured during a behavioral trial session.

From infancy, people show a preference for stimuli that exhibit certain low-level feature properties. For example, a four-month-old infant is more likely to look at a moving object than a static one, or a face-like object than one that has similar, but jumbled, features (Fagan, 1988). Both Cog and Kismet use a perceptual system which combine basic feature detectors including face detectors, motion detectors, skin color filters, and color saliency analysis. Low-level perceptual inputs are combined with high-level influences from motivations, behaviors, and habituation effects.

This system is based upon models of adult human visual search and attention (Wolfe, 1994) and has been reported previously (Breazeal & Scassellati, 1999b). The attention process constructs a linear combination of the input feature detectors and a time-decayed Gaussian field which represents habituation effects. High areas of activation in this composite generate a saccade to that location and compensatory neck movement. The weights of the feature detectors can be influenced by the motivational and behavioral state of the robot to preferentially bias certain stimuli. For example, if the robot is searching for a playmate, the weight of the face detector can be increased to cause the robot to show a preference for attending to faces. The addition of saliency cues based on the model's focus of attention can easily be incorporated into this model of attention, but the perceptual abilities needed to obtain the focus of attention have yet to be fully developed.

Title:
/home/ap/scaz/AAAI/IJCAI/matlab/seek_people.eps
Creator:
MATLAB, The Mathworks, Inc.
Preview:
This EPS picture was not saved
with a preview included in it.
Comment:
This EPS picture will print to a
PostScript printer, but not to
other types of printers.

Title:
/home/ap/scaz/AAAI/IJCAI/matlab/seek_toy.eps
Creator:
MATLAB, The Mathworks, Inc.
Preview:
This EPS picture was not saved
with a preview included in it.
Comment:
This EPS picture will print to a
PostScript printer, but not to
other types of printers.

Title:
/home/ap/scaz/AAAI/IJCAI/matlab/avoid_people.eps
Creator:
MATLAB, The Mathworks, Inc.
Preview:
This EPS picture was not saved
with a preview included in it.
Comment:
This EPS picture will print to a
PostScript printer, but not to
other types of printers.

Title:
/home/ap/scaz/AAAI/IJCAI/matlab/avoid_toy.eps
Creator:
MATLAB, The Mathworks, Inc.
Preview:
This EPS picture was not saved
with a preview included in it.
Comment:
This EPS picture will print to a
PostScript printer, but not to
other types of printers.

Fig. Preferential looking based on habituation and top-down influences. When presented with two salient stimuli (a face and a brightly colored toy), the robot prefers to look at the stimulus that has behavioral relevance to the currently active goal (shown at top). Habituation causes the robot to also spend time looking at the non-preferred stimulus.

Expressive displays: Facial expressions and body posture

By identifying the emotional states of the instructor and responding with its own emotional displays, our robots will have additional information to help determine what to imitate, evaluate success, and provide a natural interface. We have developed robots with the ability to display facial expressions and have developed emotional models that drive them based upon environmental stimuli, behavioral state, and internal motivations (Breazeal & Scassellati, 1999a). The robot's emotional responses are implemented though a variety of affective circuits, each of which implements one of the six basic emotions hypothesized to be innate in humans (anger, disgust, fear, joy, sadness, and surprise) (Ekman & Davidson, 1994). The activation of an emotional response depends upon the affective contributions that each circuit receives from drives, behaviors, and perceptual stimuli.

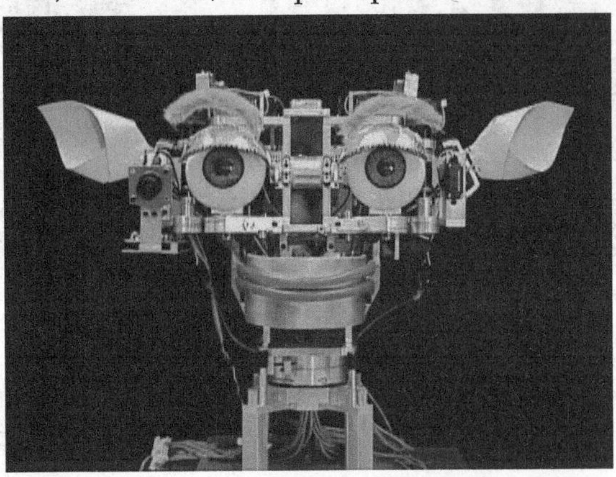

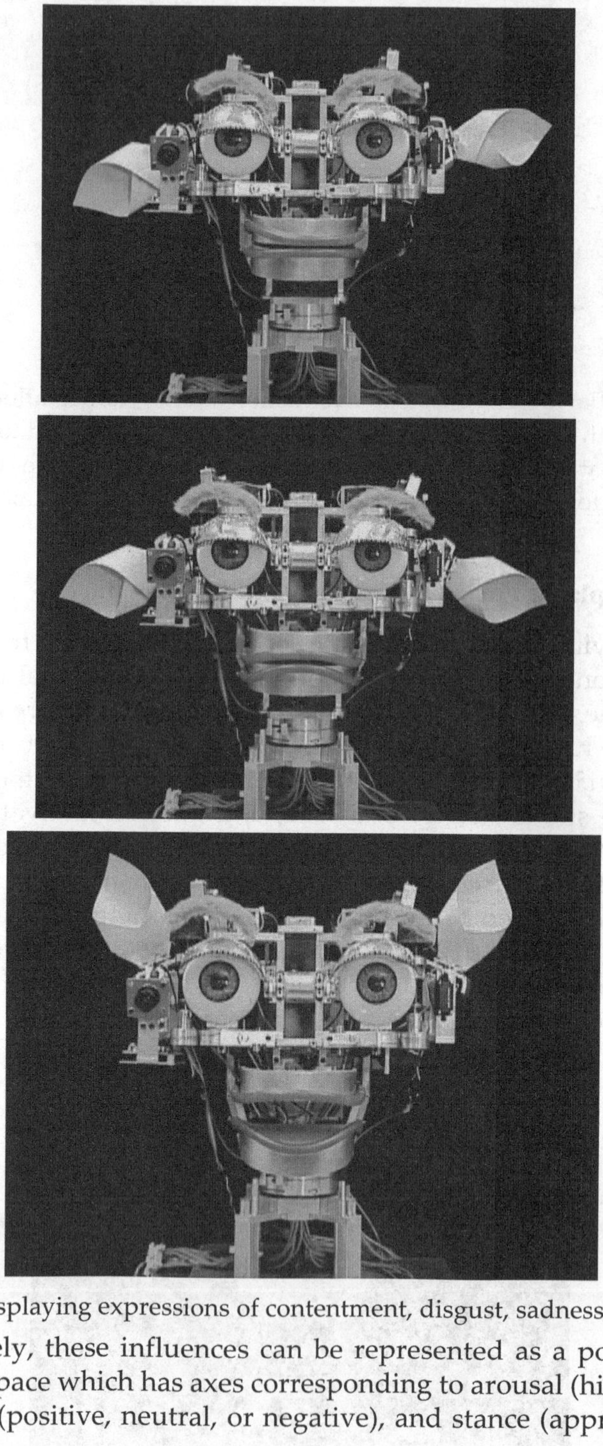

Fig. Kismet displaying expressions of contentment, disgust, sadness, and surprise.

Collectively, these influences can be represented as a point in a three dimensional space which has axes corresponding to arousal (high, neutral, or low), valence (positive, neutral, or negative), and stance (approach, neutral, or withdraw).

To generate the facial expression of the robot, each dimension of this space has a characteristic facial posture and body posture (the basis set). The resulting facial expression is an average of these basis postures weighted by the location of the affective state within this space. For example, more negatively valenced values result in having the robot frown more. The basis set of face and body postures are chosen so that each generated expression is reminiscent of the corresponding facial expression and body posture in humans when in an analogous affective state.

An initial web-based study demonstrated that both valence and arousal in the robot's expression were included in subjects' descriptions of photos of the robot, while the robot's stance was present less often in subjects' descriptions (Breazeal & Foerst, 1999).

Regulating social exchange

To learn efficiently, the robot must be capable of regulating the rate and intensity of instruction to match its current understanding and capabilities. Expressive displays combine with knowledge of social processes (such as turn taking) to allow the robot to regulate the interaction to optimize its own learning. For example, if the instructor is moving too quickly, the robot will have a difficult time maintaining the interaction and will respond with a frustrated and angry expression. In our informal observations, these behaviors are readily interpreted even by naïve instructors.

Title:
/home/mb1/ferrell/kismet/face.eps
Creator:
MATLAB, The Mathworks, Inc.
Preview:
This EPS picture was not saved
with a preview included in it.
Comment:
This EPS picture will print to a
PostScript printer, but not to
other types of printers.

Fig. This is a trace of Kismet's perceptual, behavioral, and motivational state while interacting with a person. When the face stimulus is absent, the social drive rises away from the homeostatic point causing the robot to display a sad expression, which encourages the human to engage the robot thereby restoring the drive. When the stimulus becomes too intense, the social drive drops away from the homeostatic point causing an expression of fear, which encourages the human to stop the interaction thereby restoring the drive.

With Kismet, we implemented a system which engages in a mutually regulatory interaction with a human while distinguishing between stimuli that can be influenced socially (face-like stimuli) and those that cannot (motion stimuli) (Breazeal & Scassellati, 2000). A human interacts with the robot through direct face-to-face interaction by waving a hand at the robot or by using a toy to play with the robot.

The perceptual system classifies these interactions in terms of their nature (engaging faces or playing with toys) and their quality (low intensity, good intensity, and overwhelming).

These stimuli are used by the robot to satiate its drives, each of which represents a basic "need" of the robot. (i.e., a need to be with people, a need to be played with, and a need for rest). Each drive contributes to the selection of the active behavior, which will act to either re-establish or to maintain that drive within homeostatic balance. The drives influence the affective state of the robot (contributing to a state of distress when a drive approaches a homeostatic limit or to a state of contentment as long as the drives remain within bounds). This mechanism is designed to activate emotional responses (such as fleeing to avoid a threatening stimulus) appropriate for the regulatory process.

In addition, the robot's facial expression and body posture are an external sign of the robot's internal state. Our informal observation is that naïve subjects given no instructions will adapt their behavior to maintain a happy and interested expression on the robot's face. Figure shows one example of how the robot's emotive cues are used to regulate the nature and intensity of social interaction, and how the nature of the interaction influences the robot's social drives and behavior.

ONGOING WORK

Our current work on building systems that are capable of social learning focuses on three areas: the recognition of vocal affect and communicative intent as a feedback signal for determining success, the use of joint reference skills to identify salient objects and to diagnose and correct errors, and the use of imitative games to distinguish between self and other, to distinguish between social and non-social stimuli, and to model human infant facial expression imitation.

Recognizing vocal affect and communicative intent

We are currently implementing an auditory system to enable our robots to recognize vocal affirmation, prohibition, and attentional bids while interacting with a human. By doing so, the robot will obtain natural social feedback on which of its actions have been successfully executed and which have not. Our approach is inspired by the findings of Fernald (1989), who has studied how parents convey both affective and communicative intent to infants through prosodic patterns of speech (including pitch, tempo, and tone of voice). These prosodic patterns may be universal, as infants have demonstrated the ability to recognize praise, prohibition and attentional bids even in unfamiliar languages. Similar to the work of Slaney (1998), we have used a multidimensional Gaussian mixture-model and simple acoustic measures such as pitch, energy, and cepstral coefficients to discriminate between these states on a database of infant-directed utterances. Ongoing work focuses on developing a real-time version of this system and integrating the system into social learning (Breazeal & Velasquez, 1998).

Joint Reference

While our current attention systems integrate perceptual and context-dependent saliency information, we are also constructing systems to utilize the model's focus of attention as a means of determining which actions and objects are relevant. Locating the model's focus of attention is also relevant for allowing incremental improvement. By observing the model's focus of attention while attempting a behavior, the student can gain valuable feedback on the expected action, on the possible outcomes of that action, and on possible corrective actions when an error occurs.

We have already constructed perceptual systems that allow us to detect faces, orient to the detected face, and obtain a high-resolution image of the model's eyes (Scassellati, 1998). We are currently working on utilizing information on the location of the pupil, the angle of gaze, the orientation of the head, and body posture to determine the object of the model's attention. This emphasis on joint reference is part of a larger project to build a "theory of mind" for the robot, which would allow it to attribute beliefs, desires, and intentions to the model and to imitate the goal of an action instead of the explicit action being performed (Scassellati, 1999A). Our models of joint reference are taken from developmental psychology, from animal behavior, and from studies of autism (Baron-Cohen, 1995).

Imitative Games

Imitative games can serve as a powerful motivator for young children. Our most recent work focuses on using the social context of an imitative game to allow the robot to perform two difficult perceptual tasks: distinguishing between stimuli that are socially responsive and stimuli that are unresponsive, and distinguishing between perceptual stimuli that are a result of the robot's own body and stimuli that correspond to other agents in the world. During an imitative game, the robot takes on two roles.

As the leader, the robot performs an action and looks for objects in the scene that perform a similar action soon thereafter. As the follower, the robot attempts to imitate the actions of a particular object in the world. Stimuli that respond socially and play an imitative game with the robot will allow it to be both a good follower (by performing a variety of actions which the robot can imitate) and a good leader (by imitating the robot). Static objects, such as a bookcase, will be poor followers and poor leaders; they will neither imitate the robot's actions nor perform actions which the robot can imitate. Objects that are good leaders but poor followers might be objects that are always unresponsive (such as a tree branch moving in the wind or a television) or people that are not interested in engaging the robot at the current time.

Objects that are good followers but poor leaders are likely to be self-motion (either reflections or direct perceptions of itself); a mirror image or a shadow never acts on its own, but is always a perfect follower. In this way, we can begin to see a means for classifying stimuli based on their similarity to the robot.

One difficulty in this approach is determining a matching between observed actions and the robot's own behaviors (the mapping problem). For actions like facial expressions, the robot is not capable of observing its own motor behaviors directly, and thus the mapping must either be innate or learned using an external reinforcement source.

We have proposed an implementation of Meltzoff and Moore's AIM model (1997) of human infant imitation of facial expressions and an implementation that allows the robot to learn a body mapping by observing its own reflections in a mirror (Breazeal, 1999A). This work is motivated by a belief that imitative games may play a functional role in developing an understanding of people and the development of social skills (Meltzoff, 1994, and Dautenhahn, 1994).

CHALLENGES IN BUILDING IMITATIVE ROBOTS

Researchers in robotics will recognize that there are many open and unsolved problems in our discussions. In this short section, we hope to provide to researchers outside robotics with some insight into where the difficulties in building these robots exist. From a practical perspective, building a robot is an enormous investment of time, engineering, money, and effort. Maintaining these systems can also be a frustrating and time-consuming process. Furthermore, to build all of these systems to operate in real time requires an enormous dedication to building computational architectures and optimized software.

Constructing the perceptual, motor, and cognitive skills that are necessary to begin to address the specific problems of imitation is extremely difficult. Figure shows a system architecture under development for our humanoid robots.

We are currently expanding this architecture to support imitative learning. Many of the skills to support the challenges of imitative learning are listed within the architecture, but certainly there are many skills that we have not yet begun to address. Most of the listed skills represent the work of large communities of researchers, with individual books, journals, and conferences dedicated to each. The integration of each of these components is also a challenging topic by itself. For example, representing the dynamic interaction between different behaviors or understanding the compromises involved in using many different perceptual filters presents new sets of challenges.

To begin to address the specific problems of imitation, each robotics research team must make some simplifying assumptions and trade-offs. Simplifications in the hardware design, the computational architecture, the perceptual systems, the behavioral repertoire, and cognitive abilities allow a research team to address the more complex issues without implementing complete solutions to other problems.

Each research team must be very careful to describe the assumptions that are made and the potential implications of these assumptions on the generality of their results. While these simplifications at one level are unavoidable, it is important to keep the big picture in mind.

Evaluating complex robotic systems presents another level of challenges. Most individual components can be evaluated as stand-alone modules using traditional engineering performance measures, such as comparisons against standardized data sets or considerations of optimization and efficiency. Evaluating the behavior of an integrated system using standard techniques from ethology and behavioral psychology is difficult for many reasons. First, before the complete behavior can be evaluated, all of the required system components must be implemented and integrated together.

Second, the particular assumptions used in constructing the systems may limit the types of interactions that the robot can be evaluated under. For example, limits to perception may restrict the robot to only certain limited classes of stimuli, or to stimuli that are marked in certain ways. Similarly, simplified sets of motor responses can limit the types of behavior that we can expect to observe.

Third, long-term studies of behavior are difficult because the hardware systems are fragile and constantly changing. Simply maintaining a robot at a given level of functionality requires full-time support, and few robotic systems are designed to operate for extended periods of time without human intervention.

Furthermore, because of the expenses of building a robot, each research robot is often supporting a variety of research studies, many of which are constantly altering the hardware platform. Fourth, comparing results between robots is difficult because of difference in the underlying assumptions and differences in the hardware platforms.

Despite these difficulties, we believe that the application of behavioral measurement techniques will be a critical step in the development of future robots. It is the goal of our research to achieve a level of functionality with our robots that would permit such an evaluation.

9

Neuronal Network Control of a Mobile Robot

INTRODUCTION

For the purpose of the present research, it is necessary that the disembodied cell culture is provided with embodiment, since a dissociated cell culture growing in isolation and receiving no sensory input is unlikely to develop useful operation since sensory input significantly affects neuronal connectivity and is involved in development of meaningful relationships necessary for useful processing.

In particular, the use of rodent primary dissociated cultured neuronal networks for the control of mobile 'animats' is a novel approach to discovering the computational capabilities of networks of biological neurones. An animat is any kind of physical or simulated artificial 'animal'. It is a contraction of the word anima-materials, anima most likely pertaining to the inner personality or the side of the unconscious mind of an individual, as described in Carl Jung's psychology. The word was coined by S.W. Wilson in 1991, in the first proceedings of the Simulation of Adaptive Behaviour. Thus an animat is an artificial creation that exhibits a minimum set of behaviours characteristic of biological life forms.

Typically, in vitro neuronal cultures consist of thousands of neurones, hence signals generated by them are highly variable and multi-dimensional. In order to extract from such data components/features of interest which are representative of the network's overall state, appropriate pre-processing and dimensionality reduction techniques must be applied.

The electrically-evoked and spontaneous responses of the neuronal network are coupled to the robot architecture via a machine learning interface mapping the features of interest to specific actuator commands. Mapping sensory data from the robot to a set of suitable stimulation protocols delivered to the neuronal network closes the robot-culture loop. Thus, signal processing can be broken down into two discrete areas: 'culture to robot', an output machine learning procedure processing recorded neuronal activity and 'robot to culture', an input mapping process, from sensor to stimulus.

Several animats reported in the literature have been constructed in order to investigate the control capacity of hybrid systems. Notably, Shkolnik created a very interesting control scheme for a simulated robot [1]. Two channels of a Multi-Electrode Array (MEA) were selected for stimulation. A stimulating signal consisted of a +/-600mV, 400µs biphasic pulse delivered at varying intervals. The concept of information coding is formed by testing the effect of exciting with a given time delay called the Inter-Probe Interval (IPI) between two probes i.e. electrode pulses. This technique gives rise to a characteristic response curve which forms the basis for deciding the animat's direction of movement using basic commands (forward, backward, left and right).

Other groups have used a simulated rat [2] which moved inside a four-wall environment including barrier objects, or the physical robots such as 'Koala' and 'Khepera' robots [3]. The latter were used in an embodiment experiment wherein one of the robots ('hybrid living robot - hybrot') would attempt to maintain a constant distance from the other which moved under random control. The Koala robot managed to successfully approach the Khepera and maintain a fixed distance from it.

DeMarse investigated the computational capacity of cultured networks and introduced the idea of implementing the results in a "real-life" problem, such as that of controlling a simulated aircraft's flight path (e.g. altitude and roll adjustments) [4].

On recent developments on application of learning techniques on neuronal cultures, G. Shahaf S. Marom [5] were one of the first groups to demonstrate achievement of desired discrete output computations by applying a simple form of supervised learning to dissociated randomly connected neuronal cultures. More recently, L. Bull & I. Uroukov [6] confirmed the results of Shahaf and moved a step further by successfully applying a Learning Classifier System (XCS) to manipulate cells into responding in pre-defined ways to electrical input signals. However, these both cases indicated successful results in about one third of experiments, indicating the underlying complexity of these networks. This also underlines the importance of first understanding key functional and effective connectivity features within these cultures before attempting to perform such learning schemes. As such, the analysis and modelling of culture activity and connectivity patterns throughout the culture's development is a key concern of our group and will provide important knowledge and feedback to the machine learning applications.

It is clear that even at such an early stage animat re-embodiments (real or virtual) have a prevailing role in the study of biological learning mechanisms. Our proposed physical and simulated animats provide the starting point for creating a proof-of-concept control loop around the neuronal culture and a basic platform for future more specific reinforcement learning experiments. As the fundamental problem is the coupling of the robot's goals to the culture's input output mapping the design of the animat's architecture discussed in this paper emphasises the need of flexibility and the use of machine learning techniques in search of such coupling.

The next section describes the main elements of the closed loop control of the animat and details the current system's architecture and section 3 describes the initial tests and our preliminary results. Section 4 provides a brief explanation of the Machine Learning (ML) context and finally Section 5 concludes with an overview of current progress and a discussion on planned future extensions.

CLOSING THE LOOP

Our animat system is constructed with a closed-loop, modular architecture in mind. Neuronal networks exhibit spatiotemporal patterns with millisecond precision [7], processing of which necessitates a very rapid response from neurophysiological recording and robot control systems. The software developed for this project runs on Linux-based workstations communicating over the Ethernet via fast server-client modules, thus providing the necessary speed and flexibility required when working with biological systems.

The study of cultured biological neurones has in recent years been greatly facilitated by the availability of commercially available systems such as the MEA's manufactured by MultiChannel Systems GmbH [8]. These consist of a glass culture petri-dish lined with an 8x8 array of electrodes as shown in Figure next.

A standard MEA measures 49 mm x 49 mm x 1 mm and its electrodes provide a bidirectional link between the culture and the rest of the system. The data acquisition hardware is provided by MCS and includes the head-stage (MEA connecting interface), amplifier, stimulus generator and PC data acquisition card.

We have created a modular closed loop system between a robotic platform and a neuronal network using Multi-Electrode Array (MEA) electrophysiological recording methods, allowing for bidirectional communication between the culture and the robot. The spontaneous electrochemical activity of the culture is used as an input to the robot's actuators and the robot's sensor readings are converted into stimulation signals received by the culture, effectively closing the loop. For the software-hardware/robotic framework we have selected the Miabot (Miabot Pro Research Pack [9]), a commercially available robotic platform exhibiting very accurate motor encoder precision (~0.5 mm) and speed (~3.5 m/s). Recording and stimulation hardware is provided by MultiChannel Systems MCS GmbH [8] and controlled via the open-source MEABench [10] software. We have also developed custom stimulator control software which interfaces with the commercially available stimulation hardware.

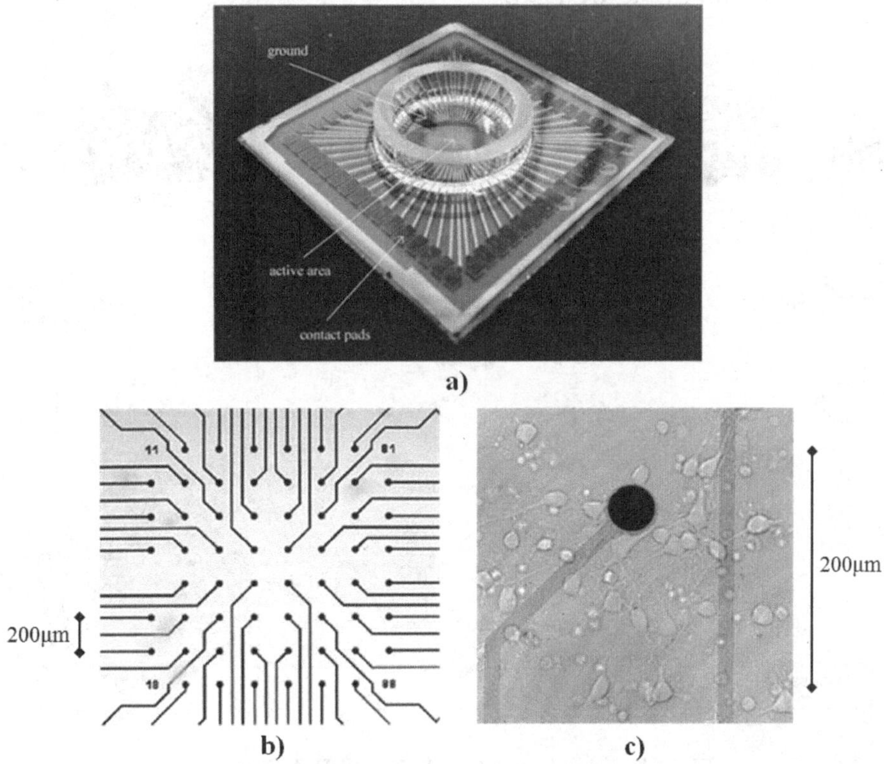

Fig. a) Typical glass MEA, showing large contact pads which lead to the electrode column – row arrangement b) Electrode arrays in the centre of the MEA, as typically seen under an optical microscope, c) Single electrode close-up, showing a number of neuronal cells in close proximity along with a vast number of neural connections between them.

A simulated counterpart for the real-life robot and its environment has also been developed, which can communicate with the culture software in the same manner as the real robot system, adding to the modular capabilities of the system. This simulation will be particularly helpful in early long-running experiments where a real robot would face issues such as power requirements, as well as in the deployment of various machine learning experiments explained below.

The closed loop system consists of several modules including the real-life or simulated robot, the MEA and stimulating hardware, a directly linked workstation for conducting computationally expensive neuronal data analyses and a separate machine running the robot control interface. The various components of the architecture communicate via TCP/IP sockets, allowing for the distribution of processing loads to multiple machines throughout the university's internal network. The modular approach to the problem can be seen in more detail in Figure next.

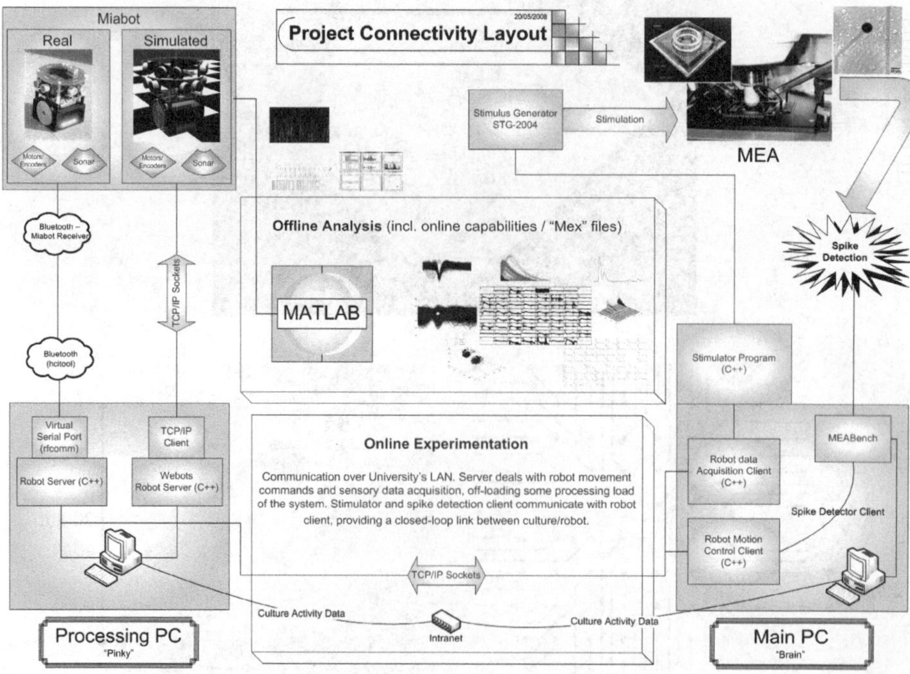

Fig. Modular Layout of Animat/MEA System

The Miabot is wirelessly controlled via Bluetooth. Communication and control is performed through custom C++ code and TCP/IP sockets, with the Server running on a PC within Bluetooth range of the Miabot, and the clients running on the main rig which has direct control of the MEA recording and stimulating software. The server sends motor commands and receives sensory data via a virtual serial port over the Bluetooth connection, while the client programs contain the closed loop code which communicates with and stimulates the MEA culture.

Subsequently, a 3D model simulation was also developed for closed loop experiments. The main advantage of any simulated device over its real-life counterpart is of course the avoidance of all potential real-life noise sources and physical constraints imposed upon the experiment. The simulation enables rapid deployment of experiments without dealing with real-time problems, power consumptions etc. An added advantage is that noise parameters can be modelled to emulate real-life conditions if required by the experiment. However, a real-life robot is the ultimate aim of the project, since its principal objective is the evaluation of the potential of live cultures in real-life problem solving situations; with all the related noise and disturbances included. The 3D Miabot model was created using the 3DStudio Max modelling package (Autodesk Inc. [11]) and exported as a Virtual Reality Modelling Language (VRML) file into the robot simulation software Webots [12], where it was further edited. A simulated environment was also created.

A custom Webots C++ controller server enables the simulated robot to either communicate via TCP/IP sockets to the same clients which perform the close loop experiments. The simulated robot server encodes and decodes information in the format of the real-life Miabot commands, thus the clients treat real-life server and simulated robot server as the same, making the whole system modular and interchangeable.

The same Webots server can also communicate data between the Matlab mathematics package [13], which is a very useful feature for the development and deployment of initial machine learning techniques.

The client code also performs text logging of all important information during the experiment run, which is then analysed offline with Matlab. Figure shows an example of collected and visualised data, such as sensory information, wheel positioning and speed, stimulator frequencies and timings, etc.

The modular approach to the architecture has resulted in a system with easily re-configurable components. The obtained closed-loop system can efficiently handle the information-rich data that is streamed via the recording software. A typical sampling frequency of 25 kHz of the culture activity recording leads to a very large required bandwidth and in consequence vast data storage space for raw data. For this reason, on-the-fly streaming of spike-detected data is the preferred method when investigating real-time closed-loop learning techniques.

A Linux-based system constitutes the core of our developed software, which also incorporates the open-source recording and data analysis toolkit MEABench, a tried-and-tested software kit used by many leading research groups around the world (Georgia Tech [14], Caltech, [15] and the University of Florida [16]).

CLOSED LOOP EXPERIMENTS

A notable recent experiment utilised the custom stimulation software in a successful closed loop test with a model cell (electronic test circuit equivalent of a passive live culture) as well as a live culture (using older control system). Initially, an appropriate neural pathway within the culture was identified and a stimulus-response electrode pair was chosen prior to the run. The pair was chosen based on the criteria that the response electrode shows minimal spontaneous activity and responds robustly to the stimulating electrode. The robot followed a forward path within its confines until it reached a wall, at which point the front sonar value dropped below a set threshold value (approx. 30 cm) which triggered a stimulation (positive-first biphasic pulse of 600 mV, 100 μs each phase). If the responding electrode registered activity (any spike) following the pulse, the robot turned in order to avoid the wall. The robot would also turn spontaneously if activity is registered on the response electrode, however the results of interest were the chain of events: Wall Detection–Stimulation–Response. Experimental data is logged during run for later analysis. The model cell experiment gives a realistic representation of how fast the closed loop response currently is, with an aim to further reduce this.

Another closed loop experiment (improved robot control system) has the robot's individual wheel speeds controlled via the frequency recorded from two chosen electrodes. This process is performed by one of the aforementioned clients, while the other client receives sonar information which is used to directly control (proportionally) the stimulating frequency of two other chosen electrodes.

The setup is reminiscent of a simple Braitenberg model, however the direct sensor-to-speed control has been interjected by the cultured network. Another group with similar research interests has previously used this method in closed loop demonstration experiments [17, 18].

The experiment above has been performed with both real and simulated robots. Run-times have only been executed for approximately 30 min., which is not enough to typically evoke Long Term Potentiation (LTP) effects between the stimulating-recording electrode sites. However, the experiment yet again was to test the performance of the improved robot control system as well as simulated robot interface, and also tested the initial data analysis tools.

This experiment has 'closed the loop' inclusive of the custom stimulation protocols and has set the basis for the following experiments which will focus on characterising the culture responses and include ML techniques for performing more complex robot control.

MACHINE LEARNING

Following the completion of the first phase of the infrastructure setup, the author's main research contribution will now lie within the application of ML techniques on the hybrid system's closed loop experiments.

These techniques may be applied in various areas, such as the spike sorting process (dimensionality reduction of spike data profiles, clustering of neuronal units), the mapping process between sensory data and culture stimulation as well as mapping between the culture activity and motor commands, and last but not least the application of learning techniques on controlled electrical stimulation of the culture, in an attempt to exploit the cultured networks' computational capacity, which have been established and demonstrated in recent research in the field.

Conclusions

The project, although still in early stages, has achieved a number of milestones and overcome a large array of technological challenges. The culture preparation techniques are constantly being refined and have lead to successful and stable cultures that exhibit both spontaneous and induced spiking/bursting activity.

A stable robotic infrastructure has been set up, tested and is in place for future machine learning and culture behaviour experiments. The animat platform consists of both a hardware element and a software simulation element. The current rate of progress could be the inspiration for offspring projects investigating culture-mediated control of a wide array of additional robotic devices, such as robotic arms/grippers, mobile robot swarms and multi-legged walkers to name but a few.

FUTURE WORK

There are a number of ways in which the current system will be expanded in the future. The Miabot may be extended to include additional sensory devices such as extra sonar arrays, mobile cameras and other range-finding hardware. A considerable current limitation is the battery power supply of an otherwise autonomous robot. A main future consideration is the inclusion of a powered-floor, which would provide the robot with relative autonomy for a longer period of time while the suggested machine learning techniques are applied and the culture's behavioural responses are monitored. The future work will adapt a Miabot to operate on an in-house powered floor, so the animat can be provided with a constant power supply; this feature is necessary since machine learning and culture behaviour tests will be carried out for many minutes and even hours at a time. At present however the robotic simulation provides an alternative solution to continuous operation of the closed loop avoiding current hardware limitations.

The current hardcoded mapping between the robot goals and the culture input output relationships will be extended to ML techniques which will reduce or even eliminate the need for a priori mapping choice. Modern RL techniques will first be applied to various mobile robot tasks such as wall following and maze navigation, in an attempt to provide a formal platform within which the actual learning capabilities of the neuronal culture will be introduced.

Also, progression of the project will require benchmarking both the machine learning techniques and the results obtained by the culture. In order to achieve this, we will aim to develop a simulation of the neural network, based on the culture's observed connectivity density and activity. This behavioural evaluation model is likely to provide great insight into the workings of the neuronal network by comparing the model's performance versus the culture's performance as well as learning capabilities as expressed by changes in its neural plasticity.

CONTROLLING A MOBILE ROBOT WITH VISUAL PROLOG

The practical realization of a simple declarative model for controlling a technical object – a mobile robot, which was built at the Robotics and Artificial Intelligence Laboratory of the State Polytechnic Museum under Technics PhD Valery Karpov and Technics PhD Dmitry Dobrynin.

Students and schoolchildren develop and program their own robots there. As an AI student I participate in this laboratory and would like to present my first robot.

ARCHITECTURE OF THE SYSTEM

The model of the robot's behavior is similar to the behavior of a living creature.

The system can be divided into the low-level actions, which are processed by the robot himself, and the high-level actions, which are suggested by a Prolog program on a stationary computer.

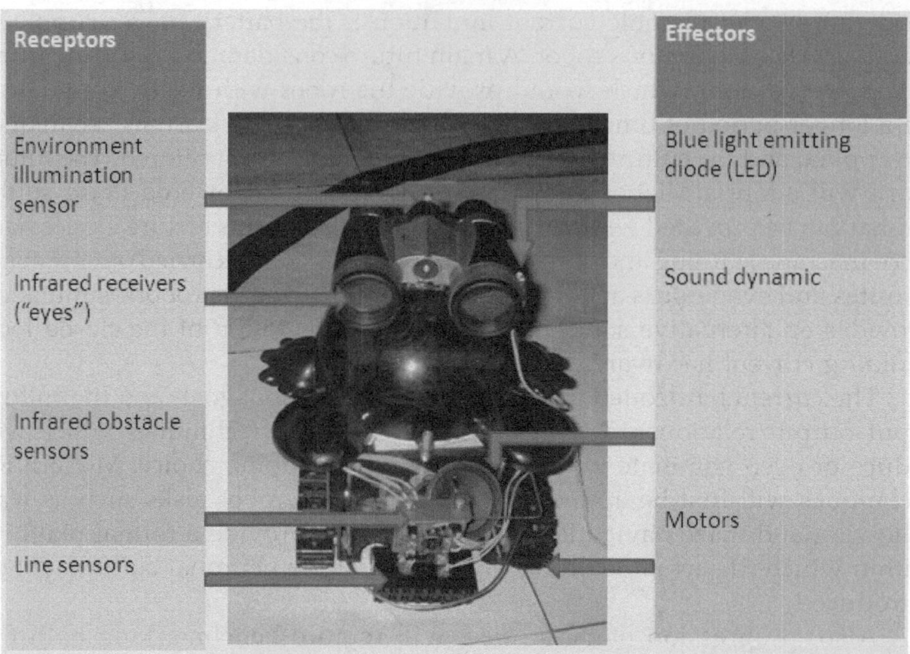

Fig. Robot's receptors and effectors

Low-level actions include: unconditional reflexes (evading an obstacle, getting frightened by a sudden light, following the path on the floor, following an IR emitter, defining whether he is at home ("home" is a grey spot on the floor)), simple actions (lighting a LED, crying), decomposition of high-level commands to low-level ones.

All low-level actions are programmed into the robot's microcomputer on C language. High-level commands are planning, guiding, database maintaining and the world map building. See Fig.2.

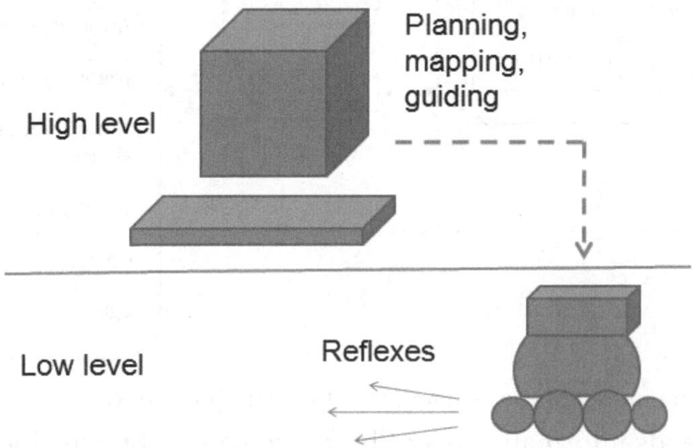

Fig. Two levels of robot's control

If robot doesn't receive any message from the computer during several seconds, he considers himself "lost" and wanders randomly, only reacting to the environmental exposure with his reflexes.

When the computer takes control, robot becomes passive and obeys the computer.

The advantage of such system is that even in the case of a mistake or a controversial command from computer, robot can still evade dangerous situations because of unconditional reflexes.

Combination of simple low-level reflexes and complex high-level command sequences gives us the model of a living organism. The idea of hierarchy in control systems is quite old, but not widely spread in robotics. Among the latest implementations, [Karpov, 2007] must be noted.

THE STRUCTURE OF THE ROBOT

The robot is an autonomous mobile device, driven by a microcontroller, which has clock rate 7 MHz and flash memory of 8 Kb, which is not enough for a complex program.

Robot has different sensors (receptors): environment illumination sensor, infrared receivers ("eyes"), infrared obstacles sensors, line photo sensors; effectors, which influence the environment: motors, light-emitting diode, sound dynamic.

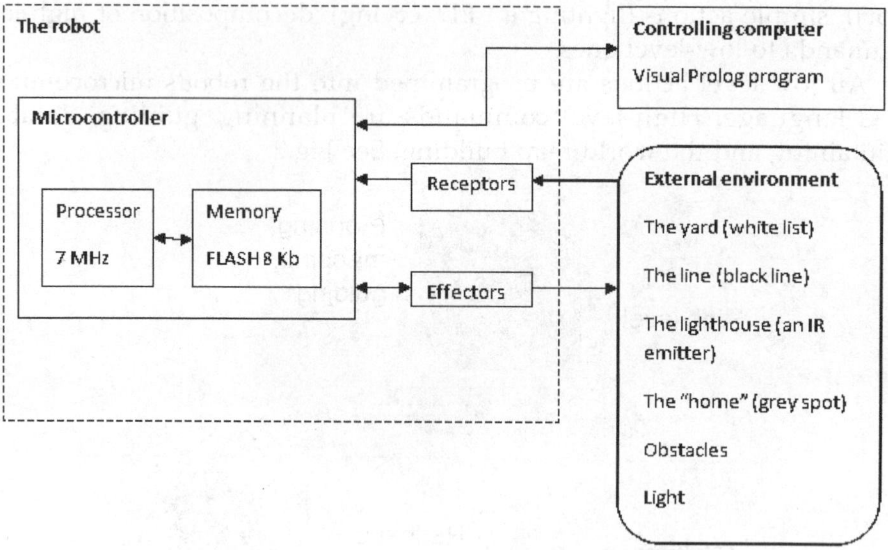

Fig. The model of the system (adopted from [Gaaze-Rapoport 1987])

Robot sends information about his movements, obtained by encoders on both driving wheels, so we can know the length of movement vector and its angle relatively to the robot's starting position. Changing polar coordinates to Cartesian, computer tries to draw robot's position on the map and assert the coordinates of all significant objects to the database to use them later for navigation.

Communication program

To establish a connection with robot, a terminal program on Visual Prolog is required. Ben Hooijenga wrote such program under VIP 7.1 CE, it's available on PDC forum and on Visual Prolog Wiki. The program is easy to apply and has a convenient GUI for setting all the parameters. The program makes it possible to send and receive bytes via RS-232 interface and process communicstion events. A declarative system for the robot was added to this program. To make a communications channel between robot and computer, a protocol was developed.

Description of the protocol

To avoid synchronization problems, a firm scheme involving timer was specified. Each 100 milliseconds a communication act takes place. Computer sends a command of sensors polling, encoded with one byte. Robot receives it and answers with two bytes, first of which is the robot's state and information about his movements. Second byte represents signals on robot's sensors. Computer decides what robot should do next, and answers with a command, typically, of one byte. See the scheme on Fig.

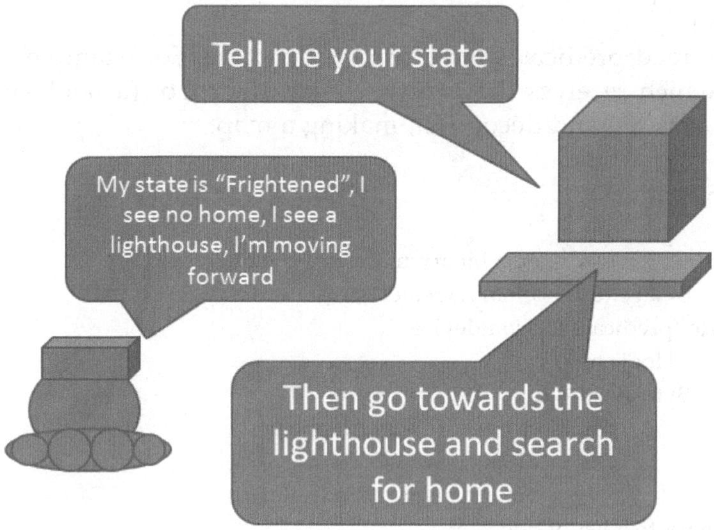

Fig. Example of communication

A fragment of the protocol on computer's side

Here is the code of timer interruption, in which communication between robot and computer takes place.

```
% timer routine on each 100 milliseconds
predicates
    onTimer: window::timerListener.
clauses
    onTimer(_Source, _timer):-
% start communication
        send("TELL_ME_YOUR_STATE"),
% get an answer from the robot about his state and movements
        readStateAndMovements(),
% get signals on sensors
        readSensors(),
% If autocontrol (base of rules) is on
    if auto_control=true() then
% get a command using a rule from base
(rule(CommandToRobot),
% send the command
            send(CommandToRobot),
        fail; % backtrack to try all the rules
        succeed())
end if,
```

```
% repaint the map on the screen
refresh_map().
```

All the read predicates have similar structure. For example, below is a predicate which receives the 2nd byte from the robot (a Boolean vector of signals on sensors) and decodes it, making a map.

```
predicates
        readSensors: ().
clauses
        readSensors():-%Create a binary massive 1 byte length
                BinaryRead = binary::create(1),
% Read 1 byte (predicate for comdevice,
written by Ben Hooijenga)
                BytesRead =
                        comdevice:readfile_overlapped(
                                BinaryRead,
                                binary::getSize(BinaryRead)),
%convert binary to a list of bits
        SensorsList = bit::bitSplit(
                binary::getIndexed_unsigned8(
                        BinaryRead, 0)),
                foreach
                        Bit = list::getMember_nd(
                                SensorsList)
                do
% decode the signals and assert the info
into database
                if decodeSensorsSignals(Bit,
                Object), ! then
% add an object, detected by sensors, to
the world model
                        placeOnTheMap(Object)
                end if
        end foreach.
```

Decoding the sensor signals simply means interpretation according to the bases of facts which are written in both robot's and computer's memory.

Here is a fragment of such a base of facts.

```
class facts
        decodeSensorsSignals: (unsigned8, string ObjectType).
clauses
        decodeSensorsSignals(128, "home").
        decodeSensorsSignals(64, "line").
        decodeSensorsSignals(32, "light").
```

And here's a fragment of base, from which commands to the robot are encoded:

```
% Base of facts - commands to the robot
% from the computer
class facts
        command: (unsigned8, string).
clauses
        command(0x10, "STOP ").
        command(0x20,
        "TELL_ME_YOUR_STATE ").
        command(0x30, "GO_FORWARD ").
```

The base of rules

The base contains rules such as:

```
predicates
        rule: (string Command)
                nondeterm (o).
clauses
% If you are frightened and see an obstacle, turn the LED on to frighten a
possible enemy
rule("SWITCH_LED_ON_TEMP "):-
        state("Frightened"),
        obstacle_sensor_active("FarIRSensor").
% If you are at home, turn your LED on
(rejoice) and wait one second (have a rest)
rule("SWITCH_LED_ON_TEMP "):-
        state("AtHome").
rule("WAIT_ONE_SECOND"):-
        state("AtHome").
% If you are still at home - go and look for line
rule("LOOK_FOR_LINE "):-
        state("AtHome").
```

An example of interaction between robot and computer

After being turned on, robot is moving around for some time to adjust his sensors. In the initial state robot moves randomly, until the computer tells him to look for the line. Having found the line, robot starts to follow it.

If a bright flash of light frightens robot, he "cries" and "panics", moving around randomly, till the computer orders him to turn LED on and follow towards the lighthouse (the lighthouse stands in the middle of "home").

Robot looks for the lighthouse and goes towards it, checking if he is already at "home". If robot reaches home, a command from computer tells him to return to the line

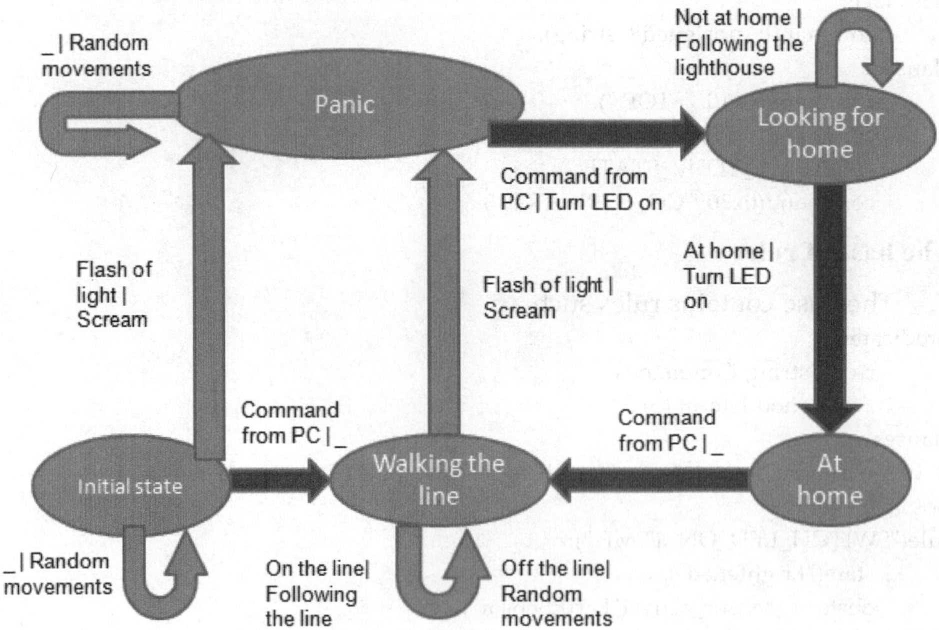

Fig. The model of robot's behavior. Red transitions can only be made after a computer's command and they do not contain in robot's memory.

Plans

Now we are working on next projects. Among our projects are continuation of work on museum robot-excursion guide and building more complicated robots. My second robot will be a robot-navigator.

SIMULATION OF ROBOT DYNAMICS AND CONTROL SOFTWARE

Most automated manufacturing tasks are done by special-purpose machines that are designed to perform prespecified functions in a manufacturing process. The inflexibility of these machines makes the computer-controlled manipulators more attractive and cost-effective in various manufacturing and assembly tasks.

Today's industrial robots, though controlled by mini-/microcomputer, are basically simple positional machines. A given task had been exacted by playing back prerecorded or preprogrammed sequences of motions that have been previously guided or taught by a user with a hand-held control/teach box.

Moreover, because the robots are equipped with few or no external sensors (both contact and noncontact), they cannot obtain vital information about their working environment. More research needs to be directed towards improving the overall performance of the manipulator systems, and one way is through the study of robot arm kinematics, dynamics, and control[Jorge Angeles 2003].

The purpose of robot arm control is to maintain the dynamic response of a computer-based manipulator in accordance with some prespecified system performance and goals. In general, the control problem consists of obtaining suitable dynamic models of the physical robot arm for designing the controller and specifying corresponding control laws or strategies to achieve the desired system response and performance. This article details the feedforward, and computed torque technique in the joint-variable space.

Simulink is a software package for modeling, simulating, and analyzing dynamic systems. It supports linear and nonlinear systems, modeled in continuous time, sampled time, or a hybrid of the two. Systems can also be multirate, i.e., have different parts that are sampled or updated at different rates.

The MatLab and Simulink environments are integrated into one entity, and thus analyzing, simulation, and revising the models can be done in either environment at any point [Steven 2006], and if Behind every robotic movement is a series of complex geometric evaluations and equations that describe motion dynamics.

Conformance to system goals must also be considered; control prototyping with Simulink allows a designer to conceptualize control solutions by means of block diagram driven environment. There is no need to use the production software and hardware during the design process thereby cutting down the cost in design. moreover, the designer can also refine the system model iteratively and tune controller parameters. these features help to reduce the implementation time, which is important in many industry situation. teaching of rapid control prototyping concept is getting acceptance by many academics in leading institutions[F C Teng 2000].

In this work the simulation of dynamic and control of robot manipulator was done by using MatLab/Simulink software; were the robotics toolbox for MatLab[Peter I. Corke 1996] was used in the modeling and simulation process. Two types of control problems was studied by using Simulink namely: feedforward control and computed torque control. to check the advantage of the of using Simulink in robot modeling and simulating the robot model chosen as industrial robot.

ROBOT DYNAMICS

Robot arm dynamics deals with the mathematical formulations of the equations of robot arm motion. The dynamic equations of manipulator motion are a set of equations describing the dynamic behavior of the manipulator.

Such equations of motion are useful for computer simulation of robot arm motion, the design of suitable control equations for a robot arm, and the evaluation of the kinematic design and structure of a robot arm.

Various approaches are available to formulate robot arm dynamics, such as the Lagrange-Euler, the Newton-Euler, the recursive Lagrange-Euler, and the generalized d'Alembert principle formulations [Spong and Vidyasagar 2001]. Deriving the dynamic model of a manipulator using the L-E method is simple and systematic. The resultant equations of motion, excluding the dynamics of the electronic control device and the gear friction, are a set of second order, coupled nonlinear differential equations.

One approach that has the advantage of both speed and accuracy is based on the N-E vector formulation[Peter I. Corke 1996] was used in this work. The derivation is simple, although messy, and involves vector cross-product terms. The resultant dynamic equations, excluding the dynamics of the control device and the gear friction, are a set of forward and backward recursive equations. These equations can be applied to the robot links sequentially. There are two problems related to manipulator dynamics that are important to solve:

- Inverse dynamics in which the manipulator's equations of motion are solved for given motion to determine the generalized forces. and
- Direct dynamics in which the equations of motion are integrated to determine the generalized coordinate response to applied generalized forces.

The equations of motion for an n-axis manipulator are given by:

$$\tau = M(q)\ddot{q} + C(q,\dot{q})\dot{q} + G(q)$$

where: if we have 4DOF

$$q = [\theta_1 \quad \theta_2 \quad \theta_3 \quad \theta_4]^T, \dot{q} = \frac{dq}{dt}, \ddot{q} = \frac{d^2q}{dt^2}$$

Obtaining the dynamic equations of motion using the MatLab program(robotics toolbox) was very powerful process; the toolbox use N-E approach to compute the equation of motion by feeding the program by necessary data about the robot system by using the functions(dyn.m, robot.m) to introduce the robot object in the MatLab program, the robot object then can be used in the MatLab program to define Simulink blocks. Because of the nature of the formulation and the method of systematically computing the torques, computations are much simpler, allowing a short computing time. With this algorithm, about three milliseconds are needed to compute the feedback joint torques per trajectory set point.

ROBOT CONTROL

Robot control is the spine of robotics. It consists in studying how to make a robot manipulator do what it is desired to do automatically; hence, it includes in designing robot controllers. Typically, these take the form of an equation or an algorithm which is realized via specialized computer programs. Then, controllers form part of the so-called robot control system which is physically constituted of a computer, a data acquisition unit, actuators (typically electrical motors), the robot itself and some extra "electronics". In this work two types of control problems was studied feedforward control and computed torque control.

Computed Torque Control

In order to overcome drawbacks of the PD controller, a more sophisticated scheme in which the magnitude of the nonlinear disturbing and loading torques is computed using the dynamic equations and used to compensate these disturbances by means of a feedforward may be employed. It must be noted that the basic control method is still PD controller with both position and velocity feedback[Rafeal and James, 1991].

The dynamic model (1) that characterizes the behavior of robot manipulators is in general, composed of nonlinear functions of the state variables (joint positions and velocities). This feature of the dynamic model might lead us to believe that given any controller, the differential equation that models the control system in closed loop should also be composed of nonlinear functions of the corresponding state variables. Nevertheless, there exists a controller which is also nonlinear in the state variables but which leads to a closed-loop control system which is described by a linear differential equation. This controller is capable of fulfilling the motion control objective.

The computed-torque control law is given by[Kelly et. al. 2005]:

$$\tau = M(q)[\ddot{q}_d + k_v \dot{\tilde{q}} + k_p \tilde{q}] + C(q,\dot{q}) + G(q)$$

where the gains k_p, k_v are chosen to meet some specific properties of the system and these gains should be adjusted to reduce the errors[Kelly and Salgado 1994]. The block diagram that corresponds to computed-torque control of robot manipulators is presented in Figure.

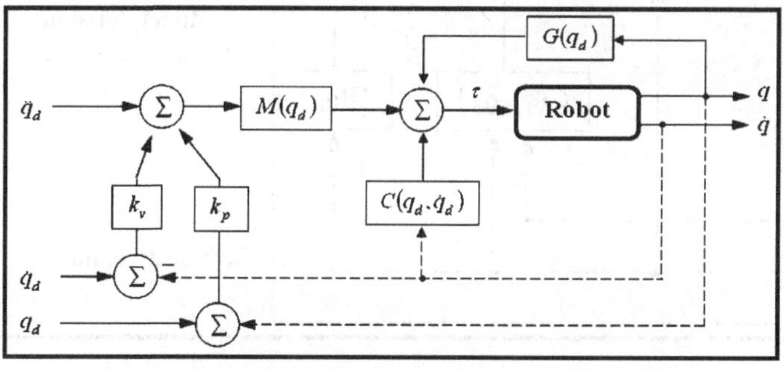

Fig. Block-diagram: computed torque control

Feedforward Control

Among the conceptually simplest control strategies that may be used to control a dynamic system we find the so-called open-loop control, where the controller is simply the inverse dynamics model of the system evaluated along the desired reference trajectories. For the case of linear dynamic systems, this control technique may be roughly presented as follows [Kelly 2005]: by using (1) applying a torque τ at the input of the robot, the behavior of its outputs q and q are governed by:

$$\frac{d}{dt}\begin{bmatrix} q \\ \dot{q} \end{bmatrix} = \begin{bmatrix} \dot{q} \\ M(q)^{-1}\left[\tau - C(q,\dot{q}) - G(q)\right] \end{bmatrix}$$

If the behavior of the outputs q and q need to be equal to that specified by q_d and q_d respectively, it seems reasonable to replace q, q and q by q_d, q_d, and q_d in the Eq. and to solve for τ. This reasoning leads to the equation of the feedforward controller, it can be expressed as:

$$\tau = M(q_d)\ddot{q}_d + C(q_d,\dot{q}_d)\dot{q}_d + G(q_d)$$

In figure the block-diagram corresponding to a robot under feedforward control is presented; the control action τ does not depend on q nor on q, that is, it is an open loop control.

The wide practical interest in incorporating the smallest number of computations in real time to implement a robot controller has been the main motivation for the PD plus feedforward control law, Feedforward control may be modified by the addition, of a feedback Proportional–Derivative (PD) term. given by:

$$\tau = M(q_d)\ddot{q}_d + C(q_d,\dot{q}_d)\dot{q}_d + G(q_d) + k_v\,\dot{\tilde{q}} + k_p\,\tilde{q}$$

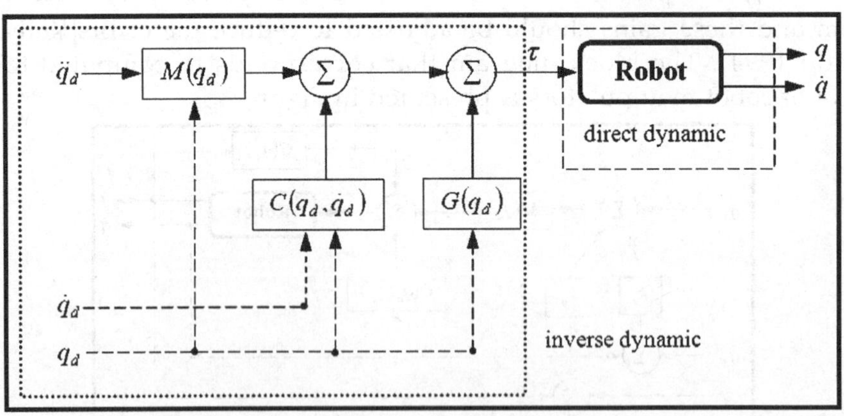

Fig. Block-diagram: feedforward control.

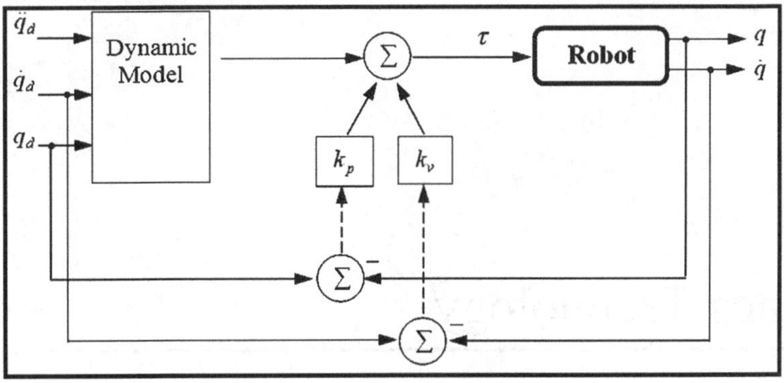

Fig. Block-diagram: PD plus feedforward control.

Simulink Model

The robot model studied in this work is the IRIS robot manipulator where all kinematic and dynamic parameters, the dynamic model for this robot system is a highly nonlinear differential equation due to the coupled between different degrees of freedom(4DOF). The modeling of the robot system was done using Robotics Toolbox in the MatLab program, using the functions(dyn.m, robot.m) to introduce the robot object in the program and then the robot object can be used with two way dynamics; The forward recursion propagates kinematics information-such as angular velocities, angular accelerations, linear accelerations, total forces and moments exerted at the center of mass of each link-from the base reference frame (inertial frame) to the end-effector. The backward recursion propagates the forces and moments exerted on each link from the end effector of the manipulator to the base reference frame. using the inverse and the direct dynamic functions which are very important in the usage of dynamic block of Simulink in the robot control simulation.

10

Robotics Technology

INTRODUCTION

Design, construction, and use of machines (robots) to perform tasks done traditionally by human beings. Robots are widely used in such industries as automobile manufacture to perform simple repetitive tasks, and in industries where work must be performed in environments hazardous to humans. Many aspects of robotics involve artificial intelligence; robots may be equipped with the equivalent of human senses such as vision, touch, and the ability to sense temperature. Some are even capable of simple decision making, and current robotics research is geared toward devising robots with a degree of self-sufficiency that will permit mobility and decision-making in an unstructured environment.

Today's industrial robots do not resemble human beings; a robot in human form is called an android. You have heard of the word "robots" during all your live; however you do not heard about the word "robotics" to often. Let star with a definition of the word itself; Robotics is a science of modern technology of general purpose of programmable machine systems. Contrary to the popular fiction image of robot as ambulatory machines of human appearance capable of performing almost any task. Most robotic systems are anchored to fixed positions in reality with limit mobility.

Robots perform a flexible, but restricted, number of operations in computer-aided manufacturing processes. These systems minimally contain a computer or a programmable device to control operations and effecters, devices that perform the desired work. The next paragraph represents the vision or general definition of robots according to the scientific knowledge and technology of that era.

ROBOTICS

Robotics is the branch of technology that deals with the design, construction, operation, and application of robots, as well as computer systems for their control, sensory feedback, and information processing. These technologies deal with automated machines that can take the place of humans in dangerous environments or manufacturing processes, or resemble humans in appearance, behavior, and/or cognition. Many of today's robots are inspired by nature contributing to the field of bio-inspired robotics. The concept of creating machines that can operate autonomously dates back to classical times, but research into the functionality and potential uses of robots did not grow substantially until the 20th century. Throughout history, robotics has been often seen to mimic human behavior, and often manage tasks in a similar fashion.

Today, robotics is a rapidly growing field, as technological advances continue, research, design, and building new robots serve various practical purposes, whether domestically,commercially, or militarily. Many robots do jobs that are hazardous to people such as defusing bombs, mines and exploring shipwrecks.

ETYMOLOGY

The word *robotics* was derived from the word *robot,* which was introduced to the public by Czech writer Karel Èapek in his play *R.U.R. (Rossum's Universal Robots),* which was published in 1920. The word *robot* comes from the Slavic word *robota,* which means labour.

The play begins in a factory that makes artificial people called *robots,* creatures who can be mistaken for humans – similar to the modern ideas of androids. Karel Èapek himself did not coin the word. He wrote a short letter in reference to an etymology in the *Oxford English Dictionary* in which he named his brother Josef Èapek as its actual originator. According to the *Oxford English Dictionary,* the word *robotics* was first used in print by Isaac Asimov, in his science fiction short story "Liar!", published in May 1941 in*Astounding Science Fiction.* Asimov was unaware that he was coining the term; since the science and technology of electrical devices is *electronics,* he assumed *robotics*already referred to the science and technology of robots. In some of Asimov's other works, he states that the first use of the word *robotics* was in his short story *Runaround*(Astounding Science Fiction, March 1942). However, the original publication of "Liar!" predates that of "Runaround" by five months, so the former is generally cited as the word's origin.

HISTORY OF ROBOTICS

In 1927 the *Maschinenmensch* ("machine-human") gynoid humanoid robot (also called "Parody", "Futura", "Robotrix", or the "Maria impersonator") was the first depiction of a robot ever to appear on film was played by German actress Brigitte Helm in Fritz Lang's film Metropolis.

In 1942 the science fiction writer Isaac Asimov formulated his Three Laws of Robotics. In 1948 Norbert Wiener formulated the principles of cybernetics, the basis of practical robotics.

Fully autonomous robots only appeared in the second half of the 20th century. The first digitally operated and programmable robot, the Unimate, was installed in 1961 to lift hot pieces of metal from a die casting machine and stack them. Commercial and industrial robots are widespread today and used to perform jobs more cheaply, or more accurately and reliably, than humans.

They are also employed in jobs which are too dirty, dangerous, or dull to be suitable for humans. Robots are widely used inmanufacturing, assembly, packing and packaging, transport, earth and space exploration, surgery, weaponry, laboratory research, safety, and the mass production of consumer and industrial goods.

COMPONENTS

Power source

At present mostly (lead-acid) batteries are used as a power source. Many different types of batteries can be used as a power source for robots. They range from lead acid batteries which are safe and have relatively long shelf lives but are rather heavy to silver cadmium batteries that are much smaller in volume and are currently much more expensive. Designing a battery powered robot needs to take into account factors such as safety, cycle lifetime and weight.

Generators, often some type of internal combustion engine, can also be used. However, such designs are often mechanically complex and need fuel, require heat dissipation and are relatively heavy. A tether connecting the robot to a power supply would remove the power supply from the robot entirely. This has the advantage of saving weight and space by moving all power generation and storage components elsewhere.

However, this design does come with the drawback of constantly having a cable connected to the robot, which can be difficult to manage.Potential power sources could be:

- Pneumatic (compressed gases)
- Hydraulics (liquids)
- Flywheel energy storage
- Organic garbage (through anaerobic digestion)
- Faeces (human, animal); may be interesting in a military context as faeces of small combat groups may be reused for the energy requirements of the robot assistant (see DEKA's project Slingshot Stirling engine on how the system would operate)

Actuation

Actuators are like the "muscles" of a robot, the parts which convert stored energy into movement. By far the most popular actuators are electric motors that spin a wheel or gear, and linear actuators that control industrial robots in factories. But there are some recent advances in alternative types of actuators, powered by electricity, chemicals, or compressed air.

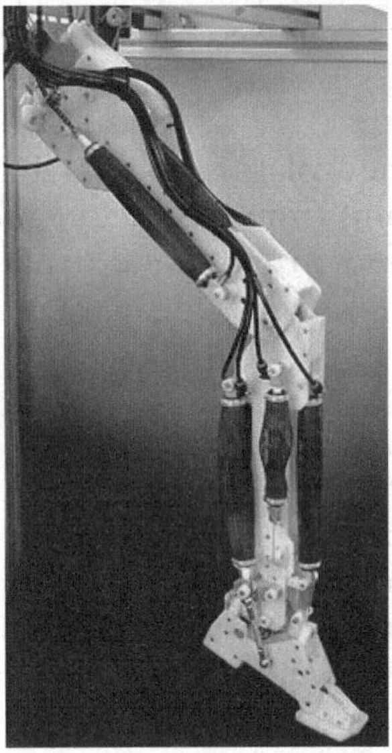

Fig. A robotic leg powered by air muscles

Electric motors

The majority of robots use electric motors, often brushed and brushless DC motors in portable robots, or AC motors in industrial robots and CNC machines. These motors are often preferred in systems with lighter loads, and where the predominant form of motion is rotational.

Linear actuators

Various types of linear actuators move in and out instead of rotating, and often have quicker direction changes, particularly when very large forces are needed such as with industrial robotics. They are typically powered by compressed air (pneumatic actuator) or an oil (hydraulic actuator).

Series elastic actuators

A spring can be designed as part of the motor actuator, to allow improved force control. It has been used in various robots, particularly walking humanoid robots.

Air muscles

Pneumatic artificial muscles, also known as air muscles, are special tubes that contract (typically up to 40%) when air is forced inside them. They have been used for some robot applications.

Muscle wire

Muscle wire, also known as Shape Memory Alloy, Nitinol or Flexinol Wire, is a material that contracts slightly (typically under 5%) when electricity runs through it. They have been used for some small robot applications.

Electroactive polymers

EAPs or EPAMs are a new plastic material that can contract substantially (up to 380% activation strain) from electricity, and have been used in facial muscles and arms of humanoid robots, and to allow new robots to float, fly, swim or walk.

Piezo motors

Recent alternatives to DC motors are piezo motors or ultrasonic motors. These work on a fundamentally different principle, whereby tiny piezoceramic elements, vibrating many thousands of times per second, cause linear or rotary motion. There are different mechanisms of operation; one type uses the vibration of the piezo elements to walk the motor in a circle or a straight line. Another type uses the piezo elements to cause a nut to vibrate and drive a screw. The advantages of these motors are nanometerresolution, speed, and available force for their size. These motors are already available commercially, and being used on some robots.

Elastic nanotubes

Elastic nanotubes are a promising artificial muscle technology in early-stage experimental development. The absence of defects in carbon nanotubes enables these filaments to deform elastically by several percent, with energy storage levels of perhaps 10 J/cm^3 for metal nanotubes. Human biceps could be replaced with an 8 mm diameter wire of this material. Such compact "muscle" might allow future robots to outrun and outjump humans.

SENSING

Sensors allow robots to receive information about a certain measurement of the environment, or internal components. This is essential for robots to perform their tasks, and act upon any changes in the environment to calculate the appropriate response. They are used for various forms of measurements, to give the robots warnings about safety or malfunctions, and to provide real time information of the task it is performing.

Touch

Current robotic and prosthetic hands receive far less tactile information than the human hand. Recent research has developed a tactile sensor array that mimics the mechanical properties and touch receptors of human fingertips. The sensor array is constructed as a rigid core surrounded by conductive fluid contained by an elastomeric skin. Electrodes are mounted on the surface of the rigid core and are connected to an impedance-measuring device within the core.

When the artificial skin touches an object the fluid path around the electrodes is deformed, producing impedance changes that map the forces received from the object. The researchers expect that an important function of such artificial fingertips will be adjusting robotic grip on held objects.

Scientists from several European countries and Israel developed a prosthetic hand in 2009, called SmartHand, which functions like a real one—allowing patients to write with it, type on a keyboard, play piano and perform other fine movements. The prosthesis has sensors which enable the patient to sense real feeling in its fingertips.

Vision

Computer vision is the science and technology of machines that see. As a scientific discipline, computer vision is concerned with the theory behind artificial systems that extract information from images. The image data can take many forms, such as video sequences and views from cameras.

In most practical computer vision applications, the computers are pre-programmed to solve a particular task, but methods based on learning are now becoming increasingly common. Computer vision systems rely on image sensors which detect electromagnetic radiation which is typically in the form of either visible light or infra-red light. The sensors are designed using solid-state physics.

The process by which light propagates and reflects off surfaces is explained using optics. Sophisticated image sensors even requirequantum mechanics to provide a complete understanding of the image formation process.

Robots can also be equipped with multiple vision sensors to be better able to compute the sense of depth in the environment. Like human eyes, robots' "eyes" must also be able to focus on a particular area of interest, and also adjust to variations in light intensities.

There is a subfield within computer vision where artificial systems are designed to mimic the processing and behavior of biological system, at different levels of complexity. Also, some of the learning-based methods developed within computer vision have their background in biology.

Other

Other common forms of sensing in robotics use LIDAR, RADAR and SONAR.

MANIPULATION

Fig. KUKA industrial robot operating in afoundry

Robots need to manipulate objects; pick up, modify, destroy, or otherwise have an effect. Thus the "hands" of a robot are often referred to as *end effectors*, while the "arm" is referred to as a *manipulator*. Most robot arms have replaceable effectors, each allowing them to perform some small range of tasks. Some have a fixed manipulator which cannot be replaced, while a few have one very general purpose manipulator, for example a humanoid hand.

For the definitive guide to all forms of robot end-effectors, their design, and usage consult the book "Robot Grippers".

Mechanical grippers

One of the most common effectors is the gripper. In its simplest manifestation it consists of just two fingers which can open and close to pick up and let go of a range of small objects. Fingers can for example be made of a chain with a metal wire run through it. Hands that resemble and work more like a human hand include the Shadow Hand, the Robonaut hand... Hands that are of a mid-level complexity include the Delft hand. An example of a simpler mechanical gripper is Cornell's universal jamming gripper, which does not use fingers but instead uses the principle of granular jamming to switch the gripper from deformable to solid.

Mechanical grippers can come in various types, including friction and encompassing jaws. Friction jaws use all the force of the gripper to hold the object in place using friction. Encompassing jaws cradle the object in place, using less friction.

Vacuum grippers

Vacuum grippers are very simple astrictive devices, but can hold very large loads provided the prehension surface is smooth enough to ensure suction. Pick and place robots for electronic components and for large objects like car screens, often use very simple vacuum grippers.

General purpose effectors

Some advanced robots are beginning to use fully humanoid hands, like the Shadow Hand, MANUS, and the Schunk hand. These are highly dexterous manipulators, with as many as 20 degrees of freedom and hundreds of tactile sensors.

LOCOMOTION

Rolling robots

For simplicity most mobile robots have four wheels or a number of continuous tracks. Some researchers have tried to create more complex wheeled robots with only one or two wheels. These can have certain advantages such as greater efficiency and reduced parts, as well as allowing a robot to navigate in confined places that a four wheeled robot would not be able to.

Two-wheeled balancing robots

Balancing robots generally use a gyroscope to detect how much a robot is falling and then drive the wheels proportionally in the opposite direction, to counterbalance the fall at hundreds of times per second, based on the dynamics of an inverted pendulum.Many different balancing robots have been designed. While the Segway is not commonly thought of as a robot, it can be thought of as a component of a robot, when used as such Segway refer to them as RMP (Robotic Mobility Platform). An example of this use has been as NASA's Robonaut that has been mounted on a Segway.

One-wheeled balancing robots

A one-wheeled balancing robot is an extension of a two-wheeled balancing robot so that it can move in any 2D direction using a round ball as its only wheel. Several one-wheeled balancing robots have been designed recently, such as Carnegie Mellon University's "Ballbot" that is the approximate height and width of a person, and Tohoku Gakuin University's "BallIP". Because of the long, thin shape and ability to maneuver in tight spaces, they have the potential to function better than other robots in environments with people.

Spherical orb robots

Several attempts have been made in robots that are completely inside a spherical ball, either by spinning a weight inside the ball, or by rotating the outer shells of the sphere. These have also been referred to as an orb bot or a ball bot.

Six-wheeled robots

Using six wheels instead of four wheels can give better traction or grip in outdoor terrain such as on rocky dirt or grass.

Tracked robots

Tank tracks provide even more traction than a six-wheeled robot. Tracked wheels behave as if they were made of hundreds of wheels, therefore are very common for outdoor and military robots, where the robot must drive on very rough terrain. However, they are difficult to use indoors such as on carpets and smooth floors. Examples include NASA's Urban Robot "Urbie".

Walking applied to robots

Walking is a difficult and dynamic problem to solve. Several robots have been made which can walk reliably on two legs, however none have yet been made which are as robust as a human. There has been much study on human inspired walking, such as AMBER lab which was established in 2008 by the Mechanical Engineering Department at Texas A&M University. Many other robots have been built that walk on more than two legs, due to these robots being significantly easier to construct. Walking robots can be used for uneven terrains, which would provide better mobility and energy efficiency than other locomotion methods. Hybrids too have been proposed in movies such as I, Robot, where they walk on 2 legs and switch to 4 (arms+legs) when going to a sprint. Typically, robots on 2 legs can walk well on flat floors and can occasionally walk up stairs. None can walk over rocky, uneven terrain. Some of the methods which have been tried are:

ZMP Technique

The Zero Moment Point (ZMP) is the algorithm used by robots such as Honda's ASIMO. The robot's onboard computer tries to keep the total inertial forces (the combination ofEarth's gravity and the acceleration and deceleration of walking), exactly opposed by the floor reaction force (the force of the floor pushing back on the robot's foot). In this way, the two forces cancel out, leaving no moment (force causing the robot to rotate and fall over). However, this is not exactly how a human walks, and the difference is obvious to human observers, some of whom have pointed out that ASIMO walks as if it needs the lavatory. ASIMO's walking algorithm is not static, and some dynamic balancing is used (see below). However, it still requires a smooth surface to walk on.

Hopping

Several robots, built in the 1980s by Marc Raibert at the MIT Leg Laboratory, successfully demonstrated very dynamic walking. Initially, a robot with only one leg, and a very small foot, could stay upright simply by hopping. The movement is the same as that of a person on a pogo stick. As the robot falls to one side, it would jump slightly in that direction, in order to catch itself. Soon, the algorithm was generalised to two and four legs. A bipedal robot was demonstrated running and even performingsomersaults. A quadruped was also demonstrated which could trot, run, pace, and bound. For a full list of these robots, see the MIT Leg Lab Robots page.

Dynamic balancing (controlled falling

A more advanced way for a robot to walk is by using a dynamic balancing algorithm, which is potentially more robust than the Zero Moment Point technique, as it constantly monitors the robot's motion, and places the feet in order to maintain stability. This technique was recently demonstrated by Anybots' Dexter Robot, which is so stable, it can even jump. Another example is the TU Delft Flame.

Passive dynamics

Perhaps the most promising approach utilizes passive dynamics where the momentum of swinging limbs is used for greater efficiency. It has been shown that totally unpowered humanoid mechanisms can walk down a gentle slope, using only gravity to propel themselves. Using this technique, a robot need only supply a small amount of motor power to walk along a flat surface or a little more to walk up a hill. This technique promises to make walking robots at least ten times more efficient than ZMP walkers, like ASIMO.

OTHER METHODS OF LOCOMOTION

Flying

A modern passenger airliner is essentially a flying robot, with two humans to manage it. The autopilot can control the plane for each stage of the journey, including takeoff, normal flight, and even landing. Other flying robots are uninhabited, and are known as unmanned aerial vehicles (UAVs). They can be smaller and lighter without a human pilot on board, and fly into dangerous territory for military surveillance missions. Some can even fire on targets under command. UAVs are also being developed which can fire on targets automatically, without the need for a command from a human. Other flying robots include cruise missiles, theEntomopter, and the Epson micro helicopter robot. Robots such as the Air Penguin, Air Ray, and Air Jelly have lighter-than-air bodies, propelled by paddles, and guided by sonar.

Snaking

Several snake robots have been successfully developed. Mimicking the way real snakes move, these robots can navigate very confined spaces, meaning they may one day be used to search for people trapped in collapsed buildings. The Japanese ACM-R5 snake robot can even navigate both on land and in water.

Skating

A small number of skating robots have been developed, one of which is a multi-mode walking and skating device. It has four legs, with unpowered wheels, which can either step or roll. Another robot, Plen, can use a miniature skateboard or roller-skates, and skate across a desktop.

Climbing

Several different approaches have been used to develop robots that have the ability to climb vertical surfaces. One approach mimics the movements of a human climber on a wall with protrusions; adjusting the center of mass and moving each limb in turn to gain leverage. An example of this is Capuchin, built by Dr. Ruixiang Zhang at Stanford University, California. Another approach uses the specialized toe pad method of wall-climbing geckoes, which can run on smooth surfaces such as vertical glass. Examples of this approach include Wallbot and Stickybot. China's "Technology Daily" November 15, 2008 reported New Concept Aircraft (ZHUHAI) Co., Ltd. Dr. Li Hiu Yeung and his research group have recently successfully developed the bionic gecko robot "Speedy Freelander". According to Dr. Li introduction, this gecko robot can rapidly climbing up and down in a variety of building walls, ground and vertical wall fissure or walking upside down on the ceiling, it is able to adapt on smooth glass, rough or sticky dust walls as well as the various surface of metallic materials and also can automatically identify obstacles, circumvent the bypass and flexible and realistic movements. Its flexibility and speed are comparable to the natural gecko. A third approach is to mimic the motion of a snake climbing a pole.

Swimming (Piscine)

It is calculated that when swimming some fish can achieve a propulsive efficiency greater than 90%. Furthermore, they can accelerate and maneuver far better than any man-made boat or submarine, and produce less noise and water disturbance. Therefore, many researchers studying underwater robots would like to copy this type of locomotion. Notable examples are the Essex University Computer Science Robotic Fish, and the Robot Tuna built by the Institute of Field Robotics, to analyze and mathematically model thunniform motion. The Aqua Penguin, designed and built by Festo of Germany, copies the streamlined shape and propulsion by front "flippers" of penguins. Festo have also built the Aqua Ray and Aqua Jelly, which emulate the locomotion of manta ray, and jellyfish, respectively.

Sailing

Sailboat robots have also been developed in order to make measurements at the surface of the ocean. A typical sailboat robot is*Vaimos* built by IFREMER and ENSTA-Bretagne. Since the propulsion of sailboat robots uses the wind, the energy of the batteries is only used for the computer, for the communication and for the actuators (to tune the rudder and the sail). If the robot is equipped with solar panel, the robot could theoretically navigate forever. The two main competitions of sailboat robots are WRSCwhich takes place every year in Europe and Sailbot.

ENVIRONMENTAL INTERACTION AND NAVIGATION

Though a significant percentage of robots in commission today are either human controlled, or operate in a static environment, there is an increasing interest in robots that can operate autonomously in a dynamic environment. These robots require some combination of navigation hardware and software in order to traverse their environment. In particular unforeseen events (e.g. people and other obstacles that are not stationary) can cause problems or collisions. Some highly advanced robots such as ASIMO, and Meinü robothave particularly good robot navigation hardware and software. Also, self-controlled cars, Ernst Dickmanns' driverless car, and the entries in the DARPA Grand Challenge, are capable of sensing the environment well and subsequently making navigational decisions based on this information. Most of these robots employ a GPS navigation device with waypoints, along with radar, sometimes combined with other sensory data such as LIDAR, video cameras, and inertial guidance systems for better navigation between waypoints.

Human-robot interaction

If robots are to work effectively in homes and other non-industrial environments, the way they are instructed to perform their jobs, and especially how they will be told to stop will be of critical importance. The people who interact with them may have little or no training in robotics, and so any interface will need to be extremely intuitive. Science fiction authors also typically assume that robots will eventually be capable of communicating with humans through speech, gestures, and facial expressions, rather than a command-line interface. Although speech would be the most natural way for the human to communicate, it is unnatural for the robot. It will probably be a long time before robots interact as naturally as the fictional C-3PO.

Speech recognition

Interpreting the continuous flow of sounds coming from a human, in real time, is a difficult task for a computer, mostly because of the great variability of speech. The same word, spoken by the same person may sound different depending on local acoustics, volume, the previous word, whether or not the speaker has a cold, etc.. It becomes even harder when the speaker has a different accent. Nevertheless, great strides have been made in the field since Davis, Biddulph, and Balashek designed the first "voice input system" which recognized "ten digits spoken by a single user with 100% accuracy" in 1952. Currently, the best systems can recognize continuous, natural speech, up to 160 words per minute, with an accuracy of 95%.

Robotic voice

Other hurdles exist when allowing the robot to use voice for interacting with humans. For social reasons, synthetic voice proves suboptimal as a communication medium,making it necessary to develop the emotional component of robotic voice through various techniques.

Gestures

One can imagine, in the future, explaining to a robot chef how to make a pastry, or asking directions from a robot police officer. In both of these cases, making hand gestureswould aid the verbal descriptions. In the first case, the robot would be recognizing gestures made by the human, and perhaps repeating them for confirmation. In the second case, the robot police officer would gesture to indicate "down the road, then turn right". It is likely that gestures will make up a part of the interaction between humans and robots.

Facial expression

Facial expressions can provide rapid feedback on the progress of a dialog between two humans, and soon may be able to do the same for humans and robots. Robotic faces have been constructed by Hanson Robotics using their elastic polymer called Frubber, allowing a large number of facial expressions due to the elasticity of the rubber facial coating and embedded subsurface motors (servos). The coating and servos are built on a metal skull. A robot should know how to approach a human, judging by their facial expression and body language. Whether the person is happy, frightened, or crazy-looking affects the type of interaction expected of the robot. Likewise, robots likeKismet and the more recent addition, Nexi can produce a range of facial expressions, allowing it to have meaningful social exchanges with humans.

Artificial emotions

Artificial emotions can also be generated, composed of a sequence of facial expressions and/or gestures. As can be seen from the movie Final Fantasy: The Spirits Within, the programming of these artificial emotions is complex and requires a large amount of human observation. To simplify this programming in the movie, presets were created together with a special software program. This decreased the amount of time needed to make the film. These presets could possibly be transferred for use in real-life robots.

Personality

Many of the robots of science fiction have a personality, something which may or may not be desirable in the commercial robots of the future. Nevertheless, researchers are trying to create robots which appear to have a personality: i.e. they use sounds, facial expressions, and body language to try to convey an internal state, which may be joy, sadness, or fear. One commercial example is Pleo, a toy robot dinosaur, which can exhibit several apparent emotions.

CONTROL

The mechanical structure of a robot must be controlled to perform tasks. The control of a robot involves three distinct phases – perception, processing, and action (robotic paradigms). Sensors give information about the environment or the robot itself (e.g. the position of its joints or its end effector). This information is then processed to calculate the appropriate signals to the actuators (motors) which move the mechanical.

The processing phase can range in complexity. At a reactive level, it may translate raw sensor information directly into actuator commands. Sensor fusion may first be used to estimate parameters of interest (e.g. the position of the robot's gripper) from noisy sensor data. An immediate task (such as moving the gripper in a certain direction) is inferred from these estimates. Techniques from control theory convert the task into commands that drive the actuators.

At longer time scales or with more sophisticated tasks, the robot may need to build and reason with a "cognitive" model. Cognitive models try to represent the robot, the world, and how they interact. Pattern recognition and computer vision can be used to track objects. Mapping techniques can be used to build maps of the world. Finally, motion planning and other artificial intelligencetechniques may be used to figure out how to act. For example, a planner may figure out how to achieve a task without hitting obstacles, falling over, etc.

Autonomy levels

Control systems may also have varying levels of autonomy.

1. Direct interaction is used for haptic or tele-operated devices, and the human has nearly complete control over the robot's motion.

2. Operator-assist modes have the operator commanding medium-to-high-level tasks, with the robot automatically figuring out how to achieve them.
3. An autonomous robot may go for extended periods of time without human interaction. Higher levels of autonomy do not necessarily require more complex cognitive capabilities. For example, robots in assembly plants are completely autonomous, but operate in a fixed pattern.

Another classification takes into account the interaction between human control and the machine motions.

1. Teleoperation. A human controls each movement, each machine actuator change is specified by the operator.
2. Supervisory. A human specifies general moves or position changes and the machine decides specific movements of its actuators.
3. Task-level autonomy. The operator specifies only the task and the robot manages itself to complete it.
4. Full autonomy. The machine will create and complete all its tasks without human interaction.

GENERAL DEFINITION FOR ROBOT

"A re-programmable, multifunctional mechanical manipulator designed to move material, parts, tools, or specialized devices through various programmed motions for the performance of a variety of tasks." – From the Robot Institute of America, 1979

WHY IS ROBOTICS IMPORTANT FOR MY CHILD?

The response is simple Robotics is a science that combines a range of fields like Mechanical Engineering, Electrical Engineering, and Computer Science. Robotics is ideal for adolescent students because it exposes them to hands-on applications of math, science, and engineering concepts. In addition, robotics motivates potential scientists and engineers to understand how things work and encourages them to use their imagination to create new technologies and improve old technologies. The next part of this background should cover the main components of a robot including some basic concepts for third to fifth grade. Now a day thinks are getting sophisticated with more technological advance. A new perception and vision of the robot representation includes the following characteristics:

Robot Components:

- Mechanical platforms – or hardware base is a mechanical device, such as a wheeled platform, arm, fixed frame or other construction, capable of interacting with its environment and any other mechanism involve with his capabilities and uses.

- Sensors systems is a special feature that rest on or around the robot. This device would be able to provide judgment to the controller with relevant information about the environment and give useful feedback to the robot.
- Joints provide more versatility to the robot itself and are not just a point that connects two links or parts that can flex, rotate, revolve and translate. Joints play a very crucial role in the ability of the robot to move in different directions providing more degree of freedom.
- The controller functions as the "brain" of the robot. Robots today have controllers that are run by programs - sets of instructions written in code. In other words, it is a computer used to command the robot memory and logic. So it, be able to work independently and automatically.
- Power Source is the main source of energy to fulfill all the robots needs. It could be a source of direct current as a battery, or alternate current from a power plant, solar energy, hydraulics or gas.
- Artificial intelligence represents the ability of computers to "think" in ways similar to human beings. Present day "AI" does allow machines to mimic certain simple human thought processes, but can not begin to match the quickness and complexity of the brain. On the other hand, not all robots possess this type of capability. It requires a lot of programming and sophisticates controllers and sensorial ability of the robot to reach this level.

The simplest behavior of a robot is locomotion. Typically, joints and wheels are used as the underlying mechanism to make a robot move from one point to the next. This type of motion should include the adaptability and versatility of the robot to continue with a specific task. *Adaptability* means adjustment to the task being carried out. *Versatility* means that the robot should have such a mechanical structure that it can carry out different tasks or perhaps the same task in different ways.

BEGINNING OF ROBOT

The word "robot" has its origin from the German word "robat". It appears that the science fiction writer Isaac Asimov was the first to use the word "robotics" to describe robot technology.

The first robots

Joseph Engel Berger, in the picture, is entitled to be the father of robotics, together with George Deroe developed the first commercial robot, *Unimate,* in 1961. It was placed on Ford and was there used for a press-loading operation. A picture of the first generation robots from Unimate can be seen in the picture below.

The first robots were principally intended to replacing humans in monotonous, heavy and hazardous processes. Distinctive features of the use of the newly developed robots were in handling of materials and work pieces without direct control or participation in the manufacturing process. Robots did not become a major force in industry generally until they had been used extensively in the Japanese automobile industry. Unimate

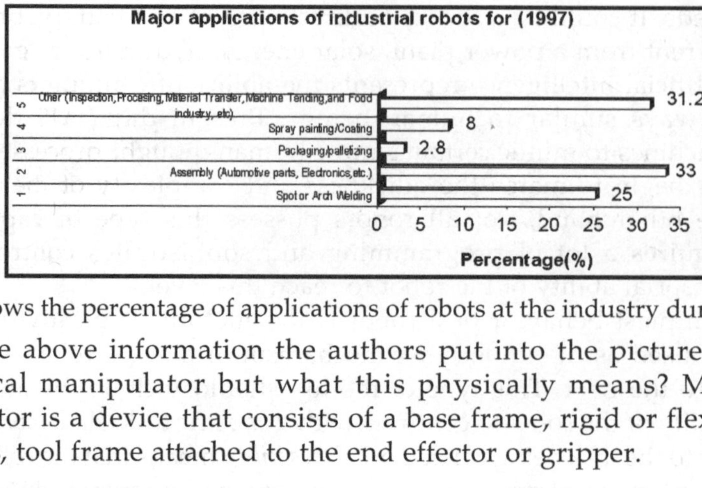

Fig. Shows the percentage of applications of robots at the industry during 1997

In the above information the authors put into the picture the word mechanical manipulator but what this physically means? Mechanical manipulator is a device that consists of a base frame, rigid or flexible links, and joints, tool frame attached to the end effector or gripper.

Physical Robot configurations

Over the years robot manufacturers have developed many types of robots of differing configurations and mechanical design, to give a variety of spatial arrangements and working volumes. These have evolved into six common types of system:

"Workspace envelope" is one of the new terms that are going to be covered in the following table. It really describes how the robot is constrained by its mechanical systems configuration. Each joint of a robot has a limit of motion range. A workspace envelope of a robot is defined as all the points in the surrounding space that can be reached by the robot. Clear under standing of the workspace envelope of a robot to be used is important because all interaction with other machines, parts, and processes only takes place within this volume of space.

Physical Configurations	
Cartesian robot it is form by 3 prismatic joints, whose axes are coincident with the X, Y and Z planes. These robots move in three directions, in translation, at right angles to each other.	SCARA robot which stands for Selective Compliance Assembly Robot Arm it is built with 2 parallel rotary joints to provide compliance in a plane. The robots work in the XY-plane and have Z-movement and a rotation of the gripper for assembly.
Cylindrical robot is able to rotate along his main axes forming a cylindrical shape. The robot arm is attached to the slide so that it can be moved radially with respect to the column.	Articulated robots are mechanic manipulator that looks like an arm with at least three rotary joints. They are used in welding and painting; gantry and conveyor systems move parts in factories.
Spherical robot is able to rotate in two different directions along his main axes and the third joint moves in translation forming a hemisphere or polar coordinate system. It used for a small number of vertical actions and is adequate for loading and unloading of a punch	Parallel robot is a complex mechanism which is constituted by two or more kinematics chains between, the base and the platform where the end-effectors are located. Good examples are the flying simulator and 4-D attractions at Univ. Studios

Index